ADVANCES IN ENGINEERING RESEARCH

ADVANCES IN ENGINEERING RESEARCH

VOLUME 25

Advances in Engineering Research

Additional books and e-books in this series can be found
on Nova's website under the Series tab.

ADVANCES IN ENGINEERING RESEARCH

ADVANCES IN ENGINEERING RESEARCH

VOLUME 25

VICTORIA M. PETROVA
EDITOR

nova
science publishers
New York

Library of Congress Cataloging-in-Publication Data

ISBN: 978-1-53614-218-1
ISSN: 2163-3932

Published by Nova Science Publishers, Inc. † New York

CONTENTS

PREFACE

Advances in Engineering Research. Volume 25 examines the background of magnetic sensors within the field of traffic engineering. Starting off with an introduction to the principles of traffic detection and estimation, the authors study the design and development of a low cost, energy efficient multi-function wireless traffic earth magnetic sensors (EMS). Recommended sensor layout strategies are presented in an effort to reduce the investment cost and provide a decision-making basis for traffic management.

Next, the authors present the main advantages of the vortex and highly turbulent flows and their implementation area in the chemical engineering. The technological scheme includes devices with active hydrodynamic regimes used in the granulation processes. The theoretical bases for the technological line of the devices with active hydrodynamic regimes is also presented.

Piezoelectric materials were discovered many years ago, however, new compounds and fabrication techniques are continually explored in order to get the optimal and new properties. The authors discuss a new piezoelectric ceramic with Platinum implant that has been fabricated in recent years. The application possibilities of this ceramic are increased since there is a third electrode available which can be used to control its behavior.

The analysis of the behaviour of Passive Magnetic Bearing (PMB) in order to achieve an acceptable magnetic force and stiffness is proposed as an interesting topic for rotating systems. In this compilation, numerical analysis, which is an effective method to investigate the structural parameters of PMB, is applied using Finite Element Method to the two-dimensional model of PMB.

The closing paper presents a universal solution for the magnetic force interaction calculation between arbitrarily magnetized permanent ring magnets. The configuration contains two permanent ring magnets that correspond to the axial or radial passive magnetic bearing configuration, and the type of the bearing depends on the permanent magnet's magnetization vector direction and the permanent magnet's position.

Chapter 1 - This chapter examines the background of magnetic sensors within the field of traffic engineering. Starting off with an introduction to the principles of traffic detection and estimation, the authors' study on the design and development of a low cost, energy efficient multi-function wireless traffic earth magnetic sensors (EMS) is then followed. With the development of sensor and network technologies, a type of hybrid sensor network has been formed in traffic information acquisition systems. Based on multi-function wireless traffic magnetic sensors and corresponding hybrid sensor network, this chapter proposes a vehicle detection algorithm based on EMS and put forwards improved vehicle detection algorithms using magnetic sensors under traffic jam conditions. The authors found that the improved vehicle detection algorithms were efficient, using accurate vehicle detection methods to address the drift problem caused by weather environments and the "tailgating effect" caused by traffic flow environments. Then, traffic state acquisition methods with different layout strategies of EMS were introduced. These methods could obtain necessary traffic information such as vehicle speed, vehicle classification and queue length for traffic state estimation. The layout strategy with a single magnetic sensor was found to be the most economical strategy to estimate queue length of the immediate past signal cycle. The experimental results also showed that the proposed single-sensor based method is simple and economical; suitable for large-scale deployment and application. Finally, sensor location problems based on fixed sensors (such as EMS) for urban road networks were studied. The heterogeneity of spatial distribution of information credibility from sensor networks was analyzed and a new measure was proposed to describe this heterogeneity and to analyze traffic spatial properties with fixed magnetic sensors. Several sensor location problem optimization models with each parameter's sensitivity analysis were proposed for urban road networks. Recommended sensor layout strategies were put forward to reduce the investment cost and to provide a decision-making basis for traffic information acquisition implementation and fine traffic management. Conclusions were given at the end.

Chapter 2 - The study presents the main advantages of the vortex and highly turbulent flows and their implementation area in the chemical engineering. The technological scheme, which includes devices with active hydrodynamic regimes used in the granulation processes, is presented in this work. It also outlines the theoretical bases for the technological line of the devices with active hydrodynamic regimes. Based on the original mathematical models, a set of applied programs is investigated to perform calculations for the granulation device. The mathematical model includes the main hydrodynamic equations for the one- and two-phase flows, thermodynamic equations, the equations of the thermal kinetics and dehydration of the granules etc. All the equations in the mathematical model are adapted for the devices with active hydrodynamic regimes. The study provides a comprehensive overview of the previous author's research results in the sphere of the automation of the granulation devices. The presented software products allow users to perform optimization calculations for the main stages of the granulation process. The obtained calculation results enable to select an optimal construction of equipment with minimal costs for the construction and provision of the technological parameters in the process.

In order to calculate the granulation and final drying stages the software products Vortex Flow©, Conical Channel©, Vortex Granulator©, Classification in vortex flow©, Multistage fluidizer© are created. The modeling of an exhaust gas cleaning process is conducted with the Vortex Tray©. All the software products allow transferring of the calculation results and making the Granulation Unit© cluster, which is demonstrated in this review for the first time.

The methodology to calculate the main operational stages of the granulation device is presented in the work. It is based on the calculation results, which were carried out with the use of the author's software products. The experimental and industrial production line of the porous ammonium nitrate, which is projected with the automated calculations by the Granulation Unit© cluster, is described.

Chapter 3 - Piezoelectric materials were discovered many years ago; however, new compounds and fabrication techniques are continually explored in order to get the optimal and new properties. In recent years, a new piezoelectric ceramic with Platinum implant has been fabricated. The application possibilities of this ceramic are increased since there is a third electrode available which can be used to control its behavior. This device is called Piezoelectric Controlled Ceramic (PCC). The PCCs were made of PLZT and $BaTi0_3$ by the oxide-mixing technique, during this process; a platinum wire of 300 μm diameter was implanted in the middle of the ceramic bulk in a transversal way. One important characteristic of the PCC is that its dielectric constant can be modified by introducing an alternate current signal in the implant. On the other hand, the response of the ceramic to light or mechanical stimuli has superior sensibility in the area near the implant. The PCC has been used in a wide variety of applications in the engineering field, such as opacity and mechanical vibration sensors. It was successfully used as a heart beat sensor (piezoplethysmography). Additionally, it has been employed as frequency modulator for wireless transmission and analog filter. The PCC with two implants has shown good results as insulation device in electrocardiography. Although the versatility of the PCC in the engineering field has been demonstrated, it is necessary to study more deeply the phenomena involved in its behavior.

Chapter 4 - The analysis of the behaviour of Passive Magnetic Bearing in order to achieve an acceptable magnetic force and stiffness is an interesting topic for rotating systems. Numerical analysis, which is an effective method to investigate the structural parameters of PMB, is applied using Finite Element Method to the two-dimensional model of

Passive Magnetic Bearing. Numerical analysis is beneficial in predicting the performances of the bearing versus different changes in the dimensions of the PMB. An optimization through Genetic Algorithms is then performed. The data gathered from the numerical analysis are therefore transferred to the Genetic Algorithm to facilitate the definition of the fitness and penalty functions which will help a faster convergence to the objective function. Providing a method to improve the magnetic force and consequently the magnetic stiffness of Passive Magnetic Bearings is an important purpose of this chapter. Also, in order to compare Passive Magnetic Bearing with different dimensions, the force to cost ratio is proposed as an index considering magnetic force and economical factors. The Genetic Algorithm is a stochastic optimization method which can be applied in order to reach the best dimensions according to the considered objective functions.

Chapter 5 - Because the magnetic bearings are contactless suspension devices, they have various rotating and translational applications. To control for the reliability of the device, it is very important to know the magnetic force interaction between the permanent magnets that are part of the device itself. This paper presents a universal solution for the magnetic force interaction calculation between arbitrarily magnetized permanent ring magnets. The configuration contains two permanent ring magnets that correspond to the axial or radial passive magnetic bearing configuration. The type of the bearing depends on the permanent magnet's magnetization vector direction and the permanent magnet's position. Calculation of the magnetic force interaction is performed when the magnetization vectors of the magnets exhibit an inclination with respect to geometry. The inclination angles of the magnetization vectors determine if the bearing will be considered radial or axial. This paper proposes a simple and fast semi-analytical approach to overcome the disadvantages of numerical approaches (e.g., high computational cost) and the disadvantages of analytical methods which are not robust enough to provide the analytical solution for force. The proposed approach is based on the magnetic charges distribution and discretization technique. The results for the magnetic force interaction obtained using the proposed approach are given for different types of the magnetic bearings and compared using the Finite Element Method (FEM). The advantage of the approach presented is its simplicity and time efficiency. The proposed approach uses only complete elliptic integrals of the second kind and additional integrations are avoided. Also, the force expression obtained allows for the analysis of different types of passive magnetic bearings.

In: Advances in Engineering Research
Editor: Victoria M. Petrova
ISBN: 978-1-53614-218-1

Chapter 1

MULTIFUNCTIONAL MAGNETIC SENSORS WITH APPLICATIONS FOR TRAFFIC MANAGEMENT

Haijian Li[1], Honghui Dong[2] and Limin Jia[2,*]
[1]Beijing Key Laboratory of Traffic Engineering,
Beijing University of Technology, Beijing, China
[2]State Key Laboratory of Rail Traffic Control and Safety,
Beijing Jiaotong University, Beijing, China

ABSTRACT

This chapter examines the background of magnetic sensors within the field of traffic engineering. Starting off with an introduction to the principles of traffic detection and estimation, our study on the design and development of a low cost, energy efficient multi-function wireless traffic earth magnetic sensors (EMS) is then followed. With the development of sensor and network technologies, a type of hybrid sensor network has been formed in traffic information acquisition systems. Based on multi-function wireless traffic magnetic sensors and corresponding hybrid

* Corresponding Author Email: jialm@vip.sina.com.

sensor network, this chapter proposes a vehicle detection algorithm based on EMS and put forwards improved vehicle detection algorithms using magnetic sensors under traffic jam conditions. We found that the improved vehicle detection algorithms were efficient, using accurate vehicle detection methods to address the drift problem caused by weather environments and the "tailgating effect" caused by traffic flow environments. Then, traffic state acquisition methods with different layout strategies of EMS were introduced. These methods could obtain necessary traffic information such as vehicle speed, vehicle classification and queue length for traffic state estimation. The layout strategy with a single magnetic sensor was found to be the most economical strategy to estimate queue length of the immediate past signal cycle. The experimental results also showed that the proposed single-sensor based method is simple and economical; suitable for large-scale deployment and application. Finally, sensor location problems based on fixed sensors (such as EMS) for urban road networks were studied. The heterogeneity of spatial distribution of information credibility from sensor networks was analyzed and a new measure was proposed to describe this heterogeneity and to analyze traffic spatial properties with fixed magnetic sensors. Several sensor location problem optimization models with each parameter's sensitivity analysis were proposed for urban road networks. Recommended sensor layout strategies were put forward to reduce the investment cost and to provide a decision-making basis for traffic information acquisition implementation and fine traffic management. Conclusions were given at the end.

Keywords: traffic management, magnetic sensors, traffic flow detection, sensor allocation

1. Introduction

1.1. Traffic State and Flow Composition

Traffic information acquisition technology provides important data for intelligent transportation systems (ITS), especially those technologies that detect road network information in points, sections or large-scale areas called sensor network technology. Becoming more prominent in many modern cities of the world, sensor network has been an important way to acquire road traffic information of ITS. For temporal and spatial data, traffic information of road sections and networks is derived from traffic data via key points in the road by making use of spatio-temporal models, traffic flow models or correlation models. In addition, due to the constraints of space-time correlation and of project investment and construction costs, it has been an important research direction of ITS to study altering methods of traffic information acquisition that use sensor network technologies (Li, 2014).

Traffic state information is an important precondition for traffic management and control and traffic information service and guidance. It is also the basis for traffic safety strategies used to detect and prevent traffic accidents. Therefore, traffic information acquisition is a significant backbone problem for ITS. Due to the application of ITS in urban traffic systems, the demand for large-scale, all-weather and all-day traffic information is high. A prior method of traffic information acquisition is traffic sensors that counts vehicles. Traffic vehicle counting from fixed sensors estimates the number of vehicles at specific locations and times in road networks. Accurate traffic vehicle counting is essential for the acquisition of other traffic information, such as traffic volumes, speed evaluation and vehicle classification. Fixed traffic sensors have inherent advantages (all-weather, all-day and fixed location) and are used for traffic information acquisition in many countries. In recent decades, many fixed devices have been proposed to obtain traffic information in the field of sensor technology, such as inductive loop detectors (Lamas-Seco et al., 2015; Oh et al., 2002), infrared instruments (Iwasaki et al., 2013), ultrasonic arrays (Jo and Jung, 2014), microwave radars (Medina et al., 2013), magnetic sensors (Cheung and Varaiya, 2007; Cheung et al., 2005; Haoui et al., 2008; Li et al., 2011; Marshall, 1978) and video image processing systems (Razakarivony and Jurie, 2016; Xia et al., 2014). However, the majority of these sensors are weather dependent or pavement unfriendly.

Nowadays, traffic problems such as traffic congestion, traffic pollution, and traffic crashes are becoming more emergent with the increase in road transportation. The advancement and application of ITS can ease these traffic problems to some extent. As one of the fundamental components of ITS, vehicle detection systems provide basic data demands. Signalized intersections are key nodes of urban road networks. Its capacity and control mechanism have a large impact on the level of service (LOS) and the connectivity of urban road networks. Queue length is the distance from the stop line to the tail of the last stopped vehicle in a single lane during red light within one signal cycle. Practically, queue length can be replaced by simply the number of vehicles stopped in a queue for convenience. Queue length plays a crucial role in signal timing optimization and LOS evaluation.

In general, loop detectors, probe vehicles, video cameras, and magnetic sensors are used to estimate vehicle queue lengths. Stable sensors, such as loop detectors, magnetic sensors, and video cameras, are typically stationed at fixed locations along roadways to record passing vehicles. When loop detectors are used alone, queue lengths can be measured by the cumulative arrivals and departures (input-output models) at an intersection; this is a common method for all stable sensors and can be used only when the queue does not exceed the locations of these stable sensors (Cai et al., 2014). Video cameras have been used to measure queue length mainly through image processing and pattern recognition. The processing algorithms include level lines (Negri, 2014) and virtual loops (Zheng et al., 2013). Queue length estimation using magnetic sensor data usually relies on a combination of measurements such as occupancy measurements at the ramp entrance, vehicle counts at the ramp's entrance and exit, speed measurements at the ramp's entrance, and vehicle identification. A type of mobile sensor, probe vehicles are used to estimate queue length based on stochastic modeling approach (Comert, 2013) or shockwave theory (Cetin, 2012). Apart from the methods that use single detection sensors, probe vehicles and loop detectors are combined together to estimate a queue length (Cai et al., 2013; Comert, 2013). Liu et al. (Liu et al., 2009) studied the real-time queue length estimation methods for congested signalized intersections using high-resolution "event-based" traffic signal data while Lighthill–Whitham–Richards (LWR) used shockwave theory to obtain a queue length estimation of the immediate past signal cycle.

Stable sensors usually have high installation and maintenance costs. Probe vehicles are typically vehicles with GPS devices. The majority of probe vehicles are taxi vehicles, which leads to a limited application for estimation of queue length when traffic flow is low. Personal mobile phones are more suitable for macroscopic traffic status estimation than microscopic queue length estimation at an intersection or a ramp. Consequently, stable sensors have distinct advantages for queue length estimation, which is used particularly for the demands of real-time traffic control and information service, or all-weather all-time detection and estimation. Magnetic sensors have an advantage over other stable sensors, such as the low installation and maintenance costs, reduction in pavement damage, less sensitive by weather challenges, and more proficient in reading lane detection and differentiation. Moreover, the advantages of magnetic sensors of being low cost, small in size, and having wireless data transmission complies with the concepts of green transportation. Using some detection algorithms, a single magnetic sensor can obtain plentiful traffic information including traffic volume, single vehicle speed and average speed, vehicle stopping and existing states, vehicle headway, vehicle classification, and so on.

1.2. Principle for Traffic Detection and Estimation

Because of the outstanding advantages and increasing popularity of magnetic sensors, the ability to count traffic vehicles by magnetic sensors has gained considerable interest. Magnetic sensors have several excellent advantages; they are low-cost, energy-efficient, small, wireless and weather-independent. Traffic surveillance by magnetic technology and corresponding wireless sensor networks has been proposed and implemented in related research areas, especially vehicle detection and counting (Cheung and Varaiya, 2007; Cheung et al., 2005; Daubaras and Zilys, 2012; Haoui et al., 2008; Li et al., 2011; Liepins and Severdaks, 2013; Marshall, 1978; Sifuentes et al., 2011; Wang et al., 2015; Yang and Lei, 2015), speed estimation (Cheung and Varaiya, 2007; Haoui et al.,

2008; Li et al., 2011) and vehicle classification (Burresi and Giorgi, 2015; Cheung and Varaiya, 2007; Cheung et al., 2005; Haoui et al., 2008; He et al., 2015; Wang et al., 2014; Yang and Lei, 2015; Zhang et al., 2010).

Early in 1978, Marshall (Marshall, 1978) proposed a magnetic-based vehicle detection method and analyzed some characteristics between magnetic fields and vehicles. Marshall noted that magnetic sensors may not be able to satisfy all requirements but should not be overlooked as a possible option for counting or sensing the presence of a vehicle. In 2007, Cheung and Varaiya (Cheung and Varaiya, 2007) investigated a type of wireless sensor network with magnetic sensors as network nodes to implement traffic surveillance. They proposed an adaptive threshold detection algorithm (ATDA) to detect vehicles. This algorithm is suitable for addressing the drift problem that is caused by temperature or light. Their experimental results are satisfactory after finding a detection accuracy of 99% from a sample size of 330 vehicles. Haoui et al. (Haoui et al., 2008) introduced the vehicle detection system to Sensys Networks Inc. and they tested this system in arterials and freeways in several U.S. cities and states. In Haoui et al. (Haoui et al., 2008), a vehicle detection system is employed for vehicle counting and for collecting vehicle occupancy and speed data. The accuracy of their system for vehicle counting and vehicle speed and occupancy data acquisition is comparable to the accuracy of well-tuned loops in different traffic congestion conditions. Sifuentes et al. (Sifuentes et al., 2011) presented a vehicle detector with a magnetic and optical sensor that was intended as a sensor node for use with a wireless sensor network (WSN). The optical sensor is used to wake-up a magnetic sensor. These two sensors, combined with energy-efficient event-based software, yielded a simple, compact, reliable, and low-energy sensor node for vehicle detection. Daubaras and Zilys (Daubaras and Zilys, 2012) introduced a vehicle detection system with magnetic field sensors and showed that the method can detect stationary vehicles in a parking lot. Liepins and Severdaks (Liepins and Severdaks, 2013) proposed a method for vehicle counting, which achieved a detection rate with an accuracy of 94.8%. The distinguishing feature of their method is the use of a type of noninvasive wireless magnetic sensor networks to eliminate the need to imbed sensors into the road structure. Wang et al. (Wang et al., 2015) employed a type of wireless magnetic sensor network to detect vehicles in an urban environment. The magnetic sensors were deployed on the sides of roads. In this paper, two wavelets methods were employed to filter noise and to improve accuracy of the magnetic signals. The accuracy of each of the two detection algorithms 90.9% and 84.1%, respectively.

In the past years, many data fusion methods have been put forward to find important traffic flow information, such as speed, flow, occupy, vehicle existence, vehicle classification, and so on. This traffic data is important to traffic control and traffic management. The experiment shows that the traffic flow, occupy and speed, all can be obtained.

Vehicle speed is one of the most important traffic parameters. Many researchers were delay based on travel time (Lin et al., 2003; Liu et al., 2012). But the method of one study to obtain the vehicle speed did not distinguish between instantaneous speed and travel speed. So how to estimate vehicle speed is an important research topic. More estimation methods using magnetic sensors can be found in papers and those methods have solved some problems for traffic surveillance (Cheung and Varaiya, 2007; Cheung et al., 2005; Kaewkamnerd et al., 2009). However, traditional methods to estimate speed may have several major drawbacks. One is that at least two sensors are always needed. Two magnetic sensors are placed separately at a distance D. When a vehicle passes over, the signal change is detected with the first sensor. After time delay Td, the signal change is detected with the second sensor. Speed v is then computed as $v = D/Td$. Another drawback is the limited placement of the sensors. Place, such as a bridge, curve, ramp exit and entrance, cannot meet the minimum interval distance requirements so these methods will not suitable. Moreover, these methods need more sensors which will increase the cost.

In recent years, ITS programs have been initiated in many countries (Lan et al., 2011). As a part of ITS, online vehicle classification system plays an important role since it gathers some critical parts of the data used in ITS. Over the past few years, methods for vehicle classification has been widely studied. Becoming an important source of information for transportation design and management, vehicle classification systems are essential for traffic management, automatic toll collection, parking optimization, and law enforcement, autonomous navigation. (Kafai and Bhanu, 2012). Moreover, transportation agencies utilize vehicle classification information to create highway maintenance plans, to evaluate highway usage and to optimize the deployment of various resources on the road (Bajwa et al., 2011). Electronic toll collection and automatic enforcement also require accurate identification of vehicle classes on highways so that vehicles of different classes are charged appropriately (Kaewkamnerd et al., 2009; Oh and Ritchie, 2007). In traffic safety research and implementation, the percentage of trucks in traffic is a critical design factor, due to their significantly lower speed and larger size (Cheung and Varaiya, 2007). In addition, vehicle classification information allows us to analyze crash data among vehicle classes, which support the derivation of more effective countermeasure for safety enhancement. In pavement design and management, pavement life is estimated according to the distribution of vehicle types running over it, and this distribution may be used to schedule re-surfacing. Furthermore, the distribution of vehicle types also provides valuable data to the prediction of highway capacity, assessment of the effectiveness of traffic legislation, weight enforcement strategies and environmental impact studies (Cheung and Varaiya, 2007). Hence, these needs have made engineers and researchers keep developing traffic surveillance systems capable of providing accurate and reliable vehicle classification information (Oh and Ritchie, 2007).

Currently, various technologies are used for vehicle classification. Loop (Coifman and Kim, 2009), video (Harlow and Peng, 2001; Kafai and Bhanu, 2012), acoustic (Paulraj et al., 2013), and other sensors (Han et al., 2011) are used in vehicle classification systems, while a large number of researchers devote themselves to the models and algorithms for vehicle classification using each of these sensors. Employing some speed estimation advances, Coifman and Kim (Coifman and Kim, 2009) obtained length-based vehicle classification using single-loop detectors. Their classification methodology was compared to concurrent measurements from video and dual-loop detectors. For video and acoustic sensors, Kafai and Bhanu (Kafai and Bhanu, 2012) presented a stochastic multiclass vehicle classification system that classifies a vehicle into one of four classes: sedan, pickup truck, SUV/minivan, and unknown. A feature set of tail light and vehicle dimensions was extracted which fed a feature selection algorithm to define a low-dimensional feature vector.

However, existing technologies such as piezoelectric sensors, inductive loops and video imaging have very high installation and maintenance costs. Some technologies such as infrared sensors and video imaging are sensitive to traffic, environmental, and weather conditions (Bajwa et al., 2011). Furthermore, the performance of microwave radar, infrared sensor and video imaging is often affected by Doppler effects or noise introduced by moving vehicles and terrain variations. In contrast, magnetic sensors are less sensitive to weather, terrain, noise and Doppler effects. Moreover, magnetic sensors have advantages in terms of cost, size, weight, power consumption and wireless communication (Nazlibilek et al., 2012). Hence, applying magnetic sensors to vehicle classification is a promising research topic.

Recently, some methods on vehicle classification or re-identification utilizing magnetic sensors have been researched (Cheung and Varaiya, 2007; Kwong et al., 2009; Lan et al., 2011). In order to improve the recognition rate by a single magnetic sensor, Lan et al. (Lan et al., 2011) proposed a novel feature extraction algorithm and a novel vehicle classification and recognition algorithm. Based on features of the concavity and convexity areas, and the angles of concave and convex parts of the waveform, they developed an improved support vector machine (ISVM) classifier to perform vehicle classification and recognition. Cheung and Varaiya (Cheung and Varaiya, 2007) studied vehicle classification using magnetic sensors based on the 13-class Federal Highway Administration (FHWA) scheme. They studied three classifiers (k-Nearest Neighbor, Support Vector Machine and Direct Hill-Pattern Matching) in different datasets. Their average accuracy rate was about 70%. If the feature of magnetic length was considered, the average accuracy rate will be slightly improved. However, in order to obtain accurate magnetic length, more magnetic sensors are needed. The existing methods are either developed with too small sample size to verify the statistical robustness or have a lack of execution time analysis which cannot reveal the effectiveness of on-line vehicle classification in magnetic sensor nodes. Besides, some methods are vehicle length-based or magnetic length-based methods which use more than one sensor to implement vehicle classification. This will greatly increase the costs of system hardware and data processing.

2. Traffic Detection Dedicated Wireless Earth Magnetic Sensors

2.1. The Earth Magnetic Sensors (EMS) Outlined

The magneto-resistance effect is a phenomenon in which the electrical resistance of a conductor or semiconductor changes in different magnetic fields. The normal magneto-resistance effect is derived from the interaction of the magnetic field and the Lorentz force of electrons. The magnetic field will cause the deflection or spiral of the movement of current carriers, which causes an increase in collision probability of electrons and enlarges the electrical resistance of a conductor or semiconductor. Several types of magneto-resistance effects, such as ordinary magneto-resistance (OMR), giant magneto-resistance (GMR) or anisotropic magneto-resistance (AMR), have been observed while AMR has been most extensively applied by researchers. Ferrum, cobalt, nickel and their respected alloys, which are each strong magnetic metals, are ferromagnetic metals. When the external magnetic field is parallel to the magnetization direction, the electrical resistance will not change with the external magnetic field. However, when the external magnetic field deviates from the magnetization direction, the electrical resistance will decrease. This phenomenon is the AMR effect of strong magnetic metals (Zhang et al., 2008). Because AMR sensors have the advantages of low energy, high sensitivity, small size, low noise, high reliability and acceptable resistance to poor environments, they are increasingly employed in several application fields.

The Honeywell HMC1001 and HMC1002 magnetic sensors are one-axis and two-axis surface mount sensors, respectively, which are designed for low-field magnetic sensing, which comprise a type of AMR sensor (Honeywell International Inc., 2008). They are extremely sensitive, low-field, solid-state magnetic sensors that are designed to measure the direction and magnitude of Earth's magnetic fields. HMC1001 is a single inline package (SIP) one-axis magnetic sensor whose sensing direction of a magnetic field is parallel to the pins; HMC1002 is a small outline-integrated circuit package (SOIC) two-axis magnetic sensor whose sensing directions are parallel and vertical to the pins. The change in the external magnetic fields will cause a change of voltage between OUT+ and OUT-, which reflects the strength of the external magnetic fields (Honeywell International Inc., 2008). The on-chip current straps are included in HMC1001/HMC1002, and external coils are not needed, which renders them easy to use.

2.2. A New Multifunctional Wireless EMS

2.2.1. Magnetic Technology-Based Traffic EMS

The multi-function traffic wireless magnetic sensor (EMS) was designed and developed by employing HMC1001/HMC1002 sensors. Traffic EMS can achieve the acquisition of several parameters of traffic flow, such as traffic volume, speed, headway and vehicle classification (Li et al., 2014a, 2011). The sensing modules of Traffic EMS are HMC1001 and HMC1002 sensors, which can detect a miniscule change in magnetic field. Because vehicles contain ferromagnetic metal, the magnetic field around the Traffic EMS will be disturbed when a vehicle passes a Traffic EMS (Bao et al., 2016b). This type of disturbance can be detected by HMC1001 and HMC1002 sensors and processed by an integrated circuit on Traffic EMS. Then, the waveform data of the vehicle, which corresponds to the disturbance, will be recorded and outputted by Traffic EMS.

By detecting the strength of a magnetic field in real time, Traffic EMS transfers the magnetic field signal to the voltage signal on an mV level and obtains the voltage signal on a V level using an amplifier. By reducing the noise signal with a filter, Traffic EMS outputs the corresponding digital value through an analog-to-digital converter (ADC). The digital value corresponds to the disturbance quantity of a magnetic field. When no vehicle exists in proximity to Traffic EMS, it will output a given digital value, which is referred to as the baseline y_b (no consideration of environmental noise). When a vehicle passes Traffic EMS, it will output waveform data with the baseline y_b and the disturbance value δ that is caused by the vehicle. The waveform data can be defined as the origin vehicle waveform data, which comprises the basis for achieving vehicle counting or other traffic information acquisition.

According to the detection and data demands, both single-module and dual-module Traffic EMS were designed and developed. Figure 1 shows the sensing axes of Traffic EMS. A single-module Traffic EMS has a HMC1001 sensor and a HMC1002 sensor. The two sensors can sense the strength of the magnetic fields of three vertical axes (X/Y/Z-axis). A dual-module Traffic EMS has two HMC1001 sensors and one HMC1002 sensor. The two HMC1001 sensors are installed by a given spacing. A dual-module Traffic EMS can sense the strength of the magnetic fields of three vertical axes (X/Y/Z1-axis) and one parallel axis (Z2-axis). In Figure 1, the X-axis points to the negative direction of running vehicles and reflects the disturbance of the magnetic field along the vehicle lane. The Y-axis points to the vertical direction of running vehicles and reflects the lateral disturbance. The Z-axis (Z1/Z2) points to the negative direction of gravity and records the presence and passing process of vehicles. The degree of sensitivity of each axis can be adjusted to adjust the valid sensing range for different applications.

Our previous study in (Li et al., 2011) presented the dual-module Traffic EMS, whose shape is a cylinder with an encircled diameter of 100 mm and a height of 65 mm. The shell material of Traffic EMS is water-, dust- and pressure-proof rigid plastic. Traffic EMS adopts a series of chips (such as CC2430/CC2530) designed by Texas Instruments to achieve energy-efficient, low-rate, low-cost, and short-range wireless communication. In practice, several wireless communication protocols (such as SimpliciTI and Zigbee) can serve as an alternative embedded in Traffic EMS to construct WSNs. These WSNs can obtain traffic flow information in a regional road network.

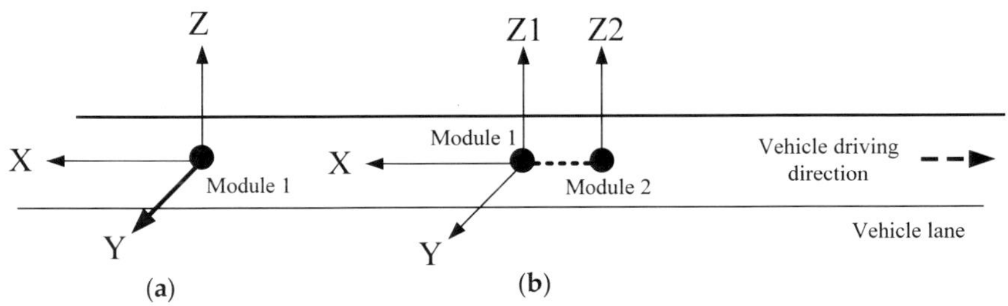

Figure 1. Sensing axes of Traffic EMS (a, single-module Traffic EMS; b, double-module Traffic EMS).

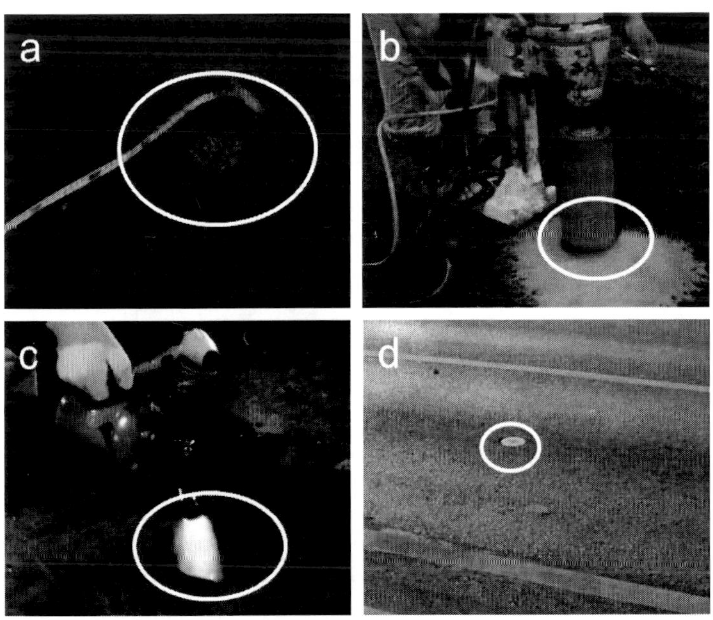

Figure 2. Quick installment of Traffic EMS (a. Selecting a position; b. Drilling a hole; c. Improving the hole; d. Installing Traffic EMS and pasting the hole).

The installation of Traffic EMS is quick and convenient due to its circular shape and technical design. Figure 2 shows the process of a four-step quick installment of Traffic EMS. The installation of Traffic EMS requires approximately 10 min, which efficiently reduces the closing time of the vehicle lanes and the disturbance for normal traffic flow.

An outstanding feature of Traffic EMS is the no-vehicle environment self-learning mechanism. After Traffic EMS is activated, the no-vehicle environment self-learning mechanism will measure and record the original strength of magnetic field when no vehicle appears around Traffic EMS. When a vehicle passes, Traffic EMS outputs a value based on the difference between the current and the original strength of the magnetic field. This mechanism will ensure that the outputs of Traffic EMS will be less influenced by weather and geographical location factors. The frequency of no-vehicle environment self-learning mechanism can be determined by setting a self-learning period in the software (Either at 1 h, 30 min, or 10 min).

2.2.2. Advantages of Traffic EMS

2.2.2.1. Low-Cost

With a reduction in the number of chips and the development of an integrated circuit, the low-cost, energy-efficient and high accurate HMC1001/HMC1002 sensors and CC2430/CC2530 communication chips are used in the integrated design of Traffic EMS. The shape of Traffic EMS is designed as a cylinder with an external diameter of only 100 mm and is encased in a water-, dust- and press- proof rigid plastic container. Thus, Traffic EMS has a remarkable economical cost compared to other traffic sensors.

2.2.2.2. Environment Self-Learning Mechanism

Through no-vehicle environment self-learning mechanism, the Traffic EMS can reduce output errors caused by temperature drift and geographical positioning. Through second traffic environment self-learning mechanism, the proposed vehicle detection algorithm can restrain the influence caused by the "tailgating effect," especially under jam flow conditions. For sectional traffic vehicle counting, which occurs under jam flow conditions, is a challenge to accurately count vehicles. These mechanisms makes it possible to obtain required detection results.

2.2.2.3. Multi-Functional Detection

Using diversified detection modules and deployment strategies, the WSNs with Traffic EMS nodes can achieve multi-parameter traffic information acquisition for the applications of traffic information service and control. A single Traffic EMS node can detect or evaluate several kinds of traffic information such as traffic volume of a lane, individual vehicle speed, time headway, and vehicle classification. One or more WSN can acquire comprehensive traffic information such as sectional traffic volume, average speed, density, occupancy and vehicle classification distribution.

2.2.2.4. Wireless Communication and Networking

Each Traffic EMS has the function of wireless communication and is able to send its detection data to roadside devices. By adopting an energy-efficient, low-rate and short-range wireless communication protocol (such as Simplitici TI and Zigbee), the Traffic EMS is able to have long-term, all-day detection. By networking several Traffic EMS nodes with roadside devices, WSNs are able to meet multiple traffic information detection demands.

2.2.2.5. Convenient Installment and Maintenance

The features of being small in size and having wireless communication capability make Traffic EMS easy to install and maintain. The installment will take less than 10 min. When the power of the Traffic EMS has been consumed, a new Traffic EMS can be reinstalled next to the original Traffic EMS; an update of the new sensor ID in the database is only necessary.

2.3. Hybrid Sensor Networks with EMS

With the development of sensor and network technology, more sensors and network devices are being deployed in road networks facilitating in ITS. The variety and hybridity of those sensors and network devices lead to a new kind of sensor network—a traffic hybrid sensor network (HSN). Knowing how to maximize the utility of HSNs in ITS has become a significant issue to study. HSNs have various patterns in term of sensor types, communication modes, transmission protocols or network structures.

HSNs can be used in various monitoring fields. Xiang et al. (Xiang et al., 2013) proposed an HSN system for indoor air quality monitoring using both stationary sensors and mobile sensors. Indoor pollutant concentration prediction models were proposed to determine the structure of the HSN. Hu et al. (W. Hu et al., 2009) studied an HSN to monitor a kind of species which required high frequency acoustic sampling, complex signal processing, wide area sensing coverage and long-lived unattended operation. In terms of smart grid monitoring, Goratti et al. (Goratti et al., 2012) proposed a novel protocol for massive scale, low-energy monitoring of critical infrastructures and used the protocol in an HSN with star and sub-mesh structure. Shenoy and Tan (Shenoy, 2005) developed a mobile robot navigation in an HSN and when an event occurred in the sensor network, a mobile robot could be navigated by the HSN to the event location.

2.3.1. Demands and Features

With the development of modern traffic information acquisition and service, no single network structure has been able to satisfy all of the multiple demands that traffic information acquisition requires. In addition, with the diversity of detection nodes and communication modes, a type of HSN has been formed in traffic information acquisition systems. Traffic information is the basis of traffic route guidance, connected vehicle systems, and automated highway systems. The multiple demands include real-time traffic volume acquisition for given road sections, traffic state detection and evaluation of key areas, vehicle detection of key locations, real-time interaction between vehicles and roadside devices, and weather detection of road environments. To meet these multiple demands, various fixed or mobile sensors and communication devices will need to be deployed on traffic roads or various other road infrastructures. Those sensors and devices will form several HSNs, which, on one hand, can be regarded as part of infrastructure of roads and urban areas, and, on the other hand, will provide the requisite information of traffic flow, infrastructure state, atmospheric conditions and environmental conditions for traffic operation management and information service systems.

The current sensor networks of traffic information acquisition in many cities are hybrid and multiple. The hybridity and diversity of sensor networks are reflected in several aspects. In terms of the detection mechanism, there are several types of detection nodes, such as loop detectors, magnetic sensors, microwave sensors, infrared sensors, and video cameras. In terms of detection objects, there are traffic flow sensors, atmospheric sensors, and infrastructure sensors. In terms of data transmission, there are text streams, audio streams, photo streams and video streams. Moreover, there are different data transmission modes and media with different communication protocols. Compared with traditional wireless sensor networks, an HSN will be influenced by the structure of the road network and the purpose of detection. For wired transmission, the network topology can be determined according to the demands of detection task and data rate, and the available communication bandwidth of each node. For wireless transmission, the optimal network topology should be determined by the spatial distributions of detection nodes, the demands of the data rate and the wireless bandwidth of communication nodes.

2.3.2. Architectures

Depending on the features and functional demands of HSNs, the definition of an HSN can be given as followed: an HSN is a set of hardware devices and software systems that are connected via wired or wireless communication media and is deployed in road networks, urban infrastructures and traffic management departments. The fundamental purpose of an HSN is to acquire various information regarding traffic flow state, infrastructure state, and atmospheric or environmental conditions. The architecture of a traffic HSN is shown in Figure 3. In the HSN, there are various devices in different layers that contain diversified traffic and environmental information with different space and time granularities.

As shown in Figure 4, in addition to end devices, an HSN requires communication nodes to serve as data transmission routes to ensure that the data detected by end devices can be transmitted to local or terminal storage devices or to the data center. The network nodes in different layers are as follows.

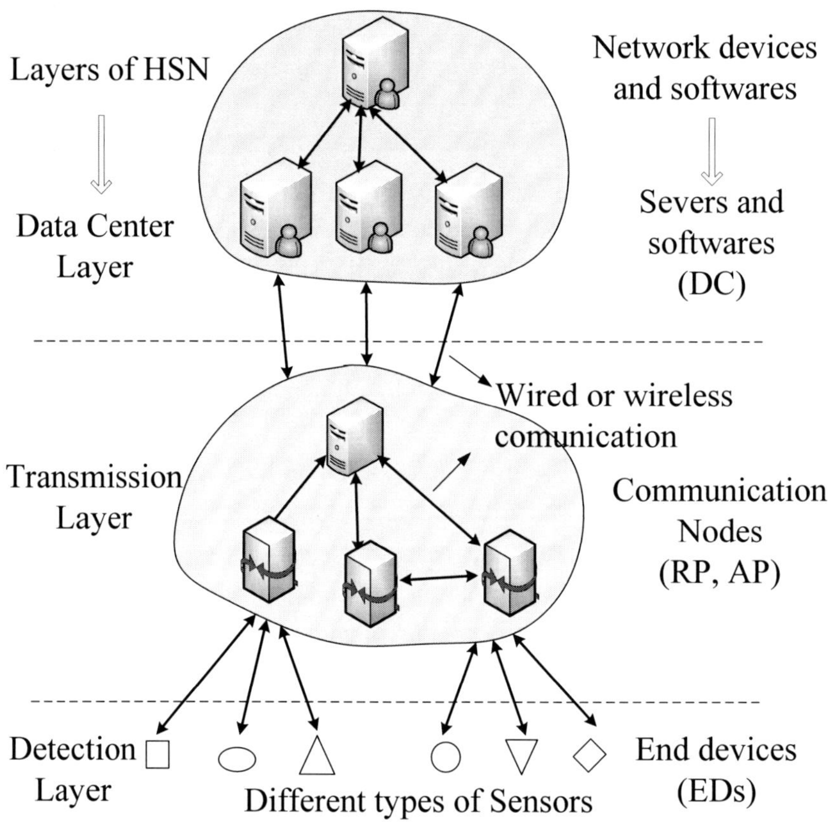

Figure 3. Architecture and devices of a traffic HSN.

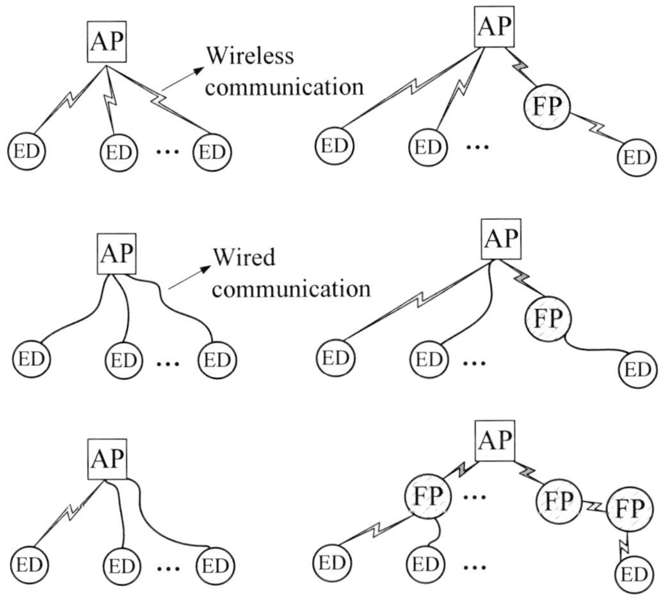

Figure 4. Alternative topologies of traffic HSNs.

1) End devices (EDs)

   EDs are the detection nodes of an HSN, such as EMS, infrastructure sensors and atmospheric sensors. As the data sources of traffic information, EDs acquire static and dynamic traffic information, which will then be transmitted to the data center via communication nodes.

2) Communication nodes: access points (APs) and repeater points (RPs)

   Communication nodes include two types of nodes in a sub-HSN. One is the RP, which extends the communication distance. In the process of wireless communication, to extend the efficient range between an ED and AP, the data from the ED is transmitted to a remote AP by multiple hop transmission. RPs utilize multiple hop transmission to implement data transmission and do not serving functions of detecting, processing or storing of information. The other type of node is the AP, which collects and processes the information, serving as the gateway of the sub-HSN while it also implements the functions of some application systems in a local area. A sub-HSN always includes one AP and up to several RPs and EDs.

3) Data Center (DC)

The DC is the center node of an HSN, which receives the information from several sub-HSNs. Serving to collect data and process information, a DC is regarded as the traffic data center of an entire city or several local areas.

According to the range of the area, a sub-HSN is built under one of two categories (Figure 4). (i) A sub-HSN includes just one AP and several EDs. (ii) In addition to the AP and ED, a sub-HSN has several RPs. For different communication modes, each category can be formed as various network topologies. A sub-HSN independently implements traffic information detection and processing in a local area. The study focuses on a single sub-HSN, and the methods of information processing and network optimization with a DC will not be discussed in the following text.

3. Traffic Flow Detection Based on EMS

Little attention has been given to vehicle detection under traffic jam conditions, especially for stop-and-go movements, which are common in cities during peak hours. When vehicles drive at low speeds in jam flow conditions, magnetic waveforms will produce a "tailgating effect" because vehicles are located close to each other. A continuous vehicle signal will be mutually interfered by front and rear vehicles, which complicates the design of a high performance vehicle detection algorithm. Thus, traffic vehicle counting in jam flow conditions will be a significant and practical topic. For low-speed congested traffic in large cities, Yang and Lei (Yang and Lei, 2015) proposed a fixed-threshold-state machine algorithm that is based on signal variance to detect vehicles within a single lane. Although the ATDA proposed in (Cheung and Varaiya, 2007) solved the drift problem, it does not consider jam flow conditions, especially the "tailgating effect" that is caused by congested traffic, and no field experiments have been conducted under jam flow conditions based on the ATDA. The traffic state under jam flow conditions is the most complicated state, since it hinders the design of an efficient and accurate vehicle detection algorithm that is based on EMS. Robust vehicle detection under jam flow conditions is challenging, especially for urban traffic.

This section introduces a multi-function wireless EMS and proposes an alternative, efficient, and accurate vehicle detection method to address the drift problem caused by weather environments and the "tailgating effect" caused by traffic flow environments. A low-cost, energy-efficient and easy installment, Traffic EMS is suitable for large-area deployment and all-weather, all-day traffic surveillance.

3.1. Vehicle Detection Based on EMS

The first step of traffic volume acquisition is single-lane vehicle detection. As a non-contact detection technology, magnetic technology has internal superiority in sectional traffic volume detection. However, the performance of a vehicle detection algorithm is related to the counting accuracy of the traffic volume, the complexity of computation and the lifetime of a sensor node.

A basic vehicle detection algorithm is appropriate when there's a long distance between the front vehicle and rear vehicle or there's a small "tailgating effect," such as during free-flow or synchronized flow conditions. Free-flow conditions always occur on a freeway, urban expressway or urban roads since there's no no influence by signalization while speeds and headways due to lack of the "tailgating effect." The traffic flow density of synchronized flow is larger and the speeds and headways are smaller than free-flow conditions. However, the "tailgating effect" can be disregarded. In Figure 5, Li et al. (Li et al., 2011) presented a vehicle waveform that is derived from Traffic EMS in free-flow conditions. Because the data of Z-axis can reflect the presence and movement of a vehicle, only the data of Z-axis will be the input of the vehicle detection algorithms in this paper, which may reduce algorithm complexity. To eliminate the "tailgating effect," the front and rear part of the vehicle waveform are almost coincident with the baseline y_b. To adopt a no-vehicle environment self-learning mechanism to reduce or to eliminate the effect of temperature or geographical position on the baseline, the baseline y_b is stable. According to the characteristics of the vehicle waveform under free-flow or synchronized flow conditions, the double window vehicle detection algorithm (Algorithm 1) that is proposed in (Li et al., 2011) is shown as follows:

Algorithm 1: double window vehicle detection algorithm
1: Initialization of the baseline y_b, the vehicle-approaching sensing window w1 and the vehicle departing sensing window w2; 2: Start the vehicle-approaching sensing window w1; 3: Input the waveform data (the input data of sampling points are the width of w1 each time); if all sampling points are within w1, go to step 4 and denote the vehicle-approaching point as P_1; otherwise, go to step 2; 4: Start the vehicle departing sensing window w2; 5: Input the waveform data (the input data of sampling points are the width of w2 each time); if all sampling points are within w2, denote the vehicle departing point as P_2 and perform vehicle counting by adding 1. The waveform data between P_1 and P_2 is referred to as the vehicle waveform; detect the next vehicle. Otherwise, go to step 4.

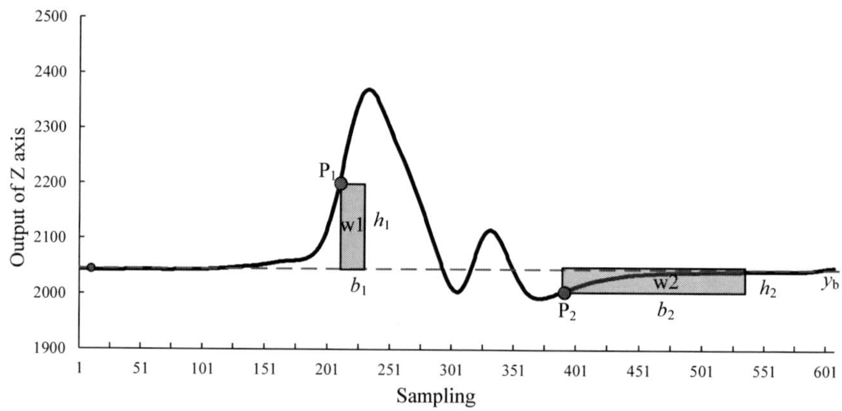

Figure 5. A vehicle waveform that is detected by Traffic EMS in free- or synchronized flow conditions.

Algorithm 1 obtains the complete waveform of each vehicle and denotes the approaching point and the departing point of each vehicle, facilitating in the acquisition of other traffic flow information, such as traffic volume, average speed, vehicle length, headway, and vehicle classification. Algorithm 1 employs the data from the Z axis as the algorithm input. The algorithm has few parameters and a high computational efficiency. Therefore, it will be easily implemented in Traffic EMS nodes with limited energy and computing capability to utilize real-time and long-time vehicle detection.

3.2. Vehicle Detection Algorithms with EMS in Traffic Jam Conditions

The jam flow primarily exists on road segments and areas with congested freeways, ramps, toll facilities, urban expressways and signalized intersections. Especially during the peak hours of urban roads, the jam flow conditions are more common. Under jam flow conditions, the density of traffic flow is large, the distance between the front vehicle and rear vehicle is small and the speed of traffic flow is slow. Sometimes, vehicles will perform stop-and-go movements or even stop and queue. Figure 6 shows the vehicle waveforms of continuous traffic flows under jam flow conditions and the "tailgating effect" between two adjacent vehicles. When Traffic EMS detects a vehicle, the front vehicle may influence the magnetic field around Traffic EMS, causing a weak performance of the Algorithm 1.

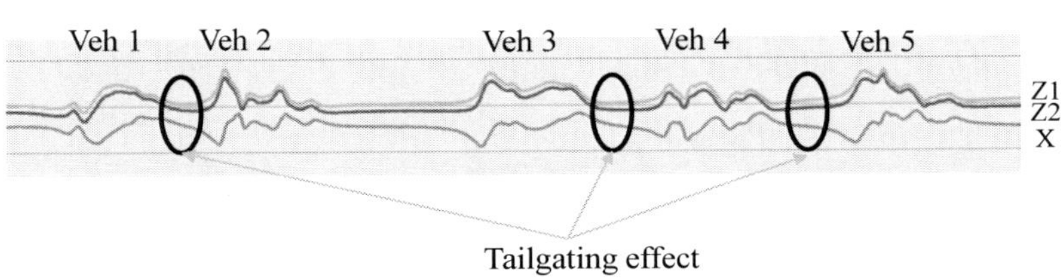

Figure 6. Waveforms of continuous traffic flow under jam flow conditions due to the tailgating effect.

Algorithm 2: second adaptive double window vehicle detection algorithm
1: Initialization of y_b, w1 and w2, and the adjacent vehicle baseline $y_{b\text{-veh}}=0$, the last baseline $y_{b\text{-old}}= y_b$, the latest baseline $y_{b\text{-new}}= y_b$, the second adaptive window w3 and the forgetting factor α.
2: Start the vehicle approaching sensing window w1 based on $y_{b\text{-new}}$;
3: Input the waveform data (the input data of sampling points are the width of w1 each time); if all sampling points are within w1, go to step 4 and denote the vehicle approaching point as p_1; otherwise, go to step 2;
4: Start the vehicle departing sensing window w2 based on $y_{b\text{-new}}$;
5: Input the waveform data (the input data of sampling points are the width of w2 each time); if all sampling points are within w2, denote the vehicle departing point as p_2 and perform vehicle counting is added by 1. If the waveform data between p_1 and p_2 is referred to as the vehicle waveform, go to step 6. Otherwise, go to step 4;
6: If all sampling points after point p_2 are within w3, denote $y_{b\text{-veh}}$ as the mean of all sampling points within w3 and go to step 7. Otherwise, go to step 2 to detect the next vehicle;
7: If $\|y_{b\text{-veh}}-y_{b\text{-old}}\|<d_{short}$ and $\|y_{b\text{-veh}}-y_b\|<d_{long}$, the update mechanism will begin; let $y_{b\text{-old}}=y_{b\text{-new}}$ and $y_{b\text{-new}}=(1-\alpha)y_{b\text{-old}}+\alpha\times y_{b\text{-veh}}$ and go to step 2; Otherwise, go to step 2 with no update operation.

In traffic engineering, a common problem is low accuracy for vehicle counting by passive invasive traffic sensors (different from active non-invasive devices, such as RFID) under jam flow conditions. Martin et al. (Martin et al., 2003) investigated the performance of different sensor technologies that are influenced by certain environmental and traffic state factors.

Although Traffic EMS has a no-vehicle environment self-learning mechanism, under jam flow conditions, the distance between the front vehicle and rear vehicle is too small to be influenced by adjacent vehicles when a vehicle passes Traffic EMS. Then, the baseline y_b of Algorithm 1 will deviate from the original value, yielding low accuracy. Due to changes of y_b, Algorithm 1 with pre-set parameters will collect only one vehicle when two or more vehicles pass Traffic EMS due to the small spacing. A simple method is to minimize the width of w2. In this manner, the probability that two or more vehicles are identified as one vehicle will be reduced; however, the probability that one vehicle is identified as two or more vehicles will increase, especially for stop-and-go vehicles and large vehicles. Based on Algorithm 1, a second adaptive double window vehicle detection algorithm (Algorithm 2) was proposed to resolve this issue. Similar to the ATDA that is proposed in (Cheung and Varaiya, 2007), Algorithm 2 realized the second self-learning for traffic environment via an updated mechanism of the baseline y_b. The first self-learning mechanism causes Traffic EMS to adapt a no-vehicle environment, whereas the second self-learning mechanism enables Traffic EMS to adapt a traffic environment. The Algorithm 2 and its traffic environment update mechanism is shown above.

For Algorithm 2, w3 is the second adaptive window, which represents the minimal "tailgating" distance between the front vehicles and rear vehicles. If the waveform data do not satisfy the criteria for w3, the subsequent waveform data represents the same vehicle with previous waveform data. The parameter α is the forgetting factor. The larger the value of α, the greater is the amount that the algorithm forgets, lowering the weight of the old baseline when calculating the new baseline. d_{short} is the short-term drift measure of y_b of Traffic EMS, which is the difference between the adjacent vehicle baseline and the last baseline. d_{long} is the long-term drift measure of y_b, that is, the difference between the adjacent vehicle baseline and the baseline. When both values of d_{short} and d_{long} are within an established range, the update mechanism of the algorithm will be activated, which can improve the stability and robustness of Algorithm 2. The parameters of Algorithm 2 can be calibrated based on field waveform data.

3.3. Experiment

To verify the detection performance of Traffic EMS and the effectiveness of the proposed vehicle detection algorithms, the Traffic EMS nodes were deployed in field traffic environments. These traffic scenarios include different traffic states of free-flow, synchronized flow and jam flow conditions. The video cameras were used to record the actual traffic flow state, and the actual traffic volume data were obtained by manual statistics based on the video data that will be employed for comparison and analysis.

3.3.1. Scenario I: Including Synchronized Flow and Jam Flow Conditions

In this scenario, Traffic EMS was deployed in the center of a lane at a sub-arterial road in Beijing. Traffic EMS can detect the real-time traffic state of this lane. This road is a two-lane road with has no disturbances, such as no lane changing among the vehicles on the road. The average speed of the vehicles on this road is a medium speed. The approximate distance between Traffic EMS and the pedestrian signal is 30 m. When the signal is green, the synchronized flow will be the dominating mode; when the signal is red, the vehicles will queue and the dominating mode of traffic flow will transfer to jam flow. Several data sets with different time periods were obtained; both Algorithm 1 and Algorithm 2 were tested based on these data sets. Table 1 shows the results of experiment scenario I for traffic vehicle counting (Bao et al., 2016b). In Table 1, each data set was named according to the starting time to be recorded, with a format of year-month-day-hour-minute.

As shown in Table 1, the performance of both Algorithm 1 and Algorithm 2 is acceptable for vehicle counting in scenario I and the mean absolute percent error (MAPE) of their accuracies exceeds 99%. In the red signal interval, especially during the evening peak hour (such as the date set 200907291708), the mode of traffic flow will convert to jam flow mode and the car-following distance between a front vehicle and a rear vehicle is very small (approximately 2 m). To reduce the disturbance of the "tailgating effect" between front and rear vehicles under jam flow conditions, Algorithm 2 may yield a better performance. For other conditions, Algorithm 1 and Algorithm 2 can each achieve satisfactory results. Although these data sets contain different climate and temperature conditions (summer and winter), different time periods (nonpeak hours and peak hours), different traffic signal conditions (smooth traffic during green signal intervals and queuing traffic during red signal intervals), the results are not significantly different under each of these conditions, indicating that Algorithm 1 and Algorithm 2 are each robust.

3.3.2. Scenario II: Including Free-Flow, Synchronized Flow and Jam Flow Conditions

This scenario is located at a signalized intersection an arterial road in Xi'an, China. Traffic EMS was deployed in the center of an outer straight lane near the signalized intersection. The traffic flow in the lane will be influenced by the traffic signal, which will cause a long vehicle queue. The vehicles in the traffic flow will exhibit car-following behaviors and stop-and-go movements due to the influence of heavy traffic. The algorithms of Algorithm 1 and Algorithm 2 were tested based on the continuous vehicle waveform data that were derived from Traffic EMS between 3 AM on November 24, 2012 and 3 AM on November 25, 2012. Figure 7 shows the traffic state within this 24-hour period. Prior to dawn, the traffic volume was low and the free-flow mode was the dominating mode. Early in the morning as the traffic volume increased, the synchronized flow mode became the dominating mode. However, during the morning or evening peak hours, the traffic volume was high and the traffic congestion increased. Then, the phenomena of stop-and-go movements frequently occurred and the jam flow mode became a main mode.

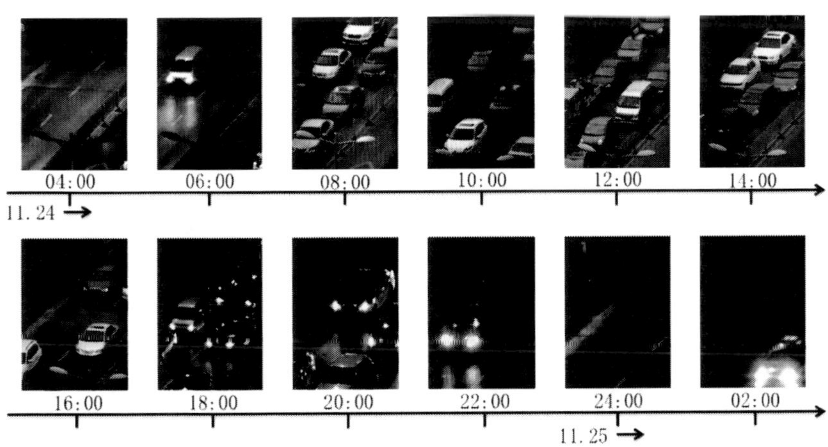

Figure 7. Field traffic flow states of Experiment scenario II.

Table 1. Results of experiment scenario I for traffic vehicle counting

Data sets	Dominant mode	Counting from Camera (Vehs)	Output of Algorithm 1 (Vehs)	Output of Algorithm 2 (Vehs)	APE of Algorithm 1 (%)	APE of Algorithm 2 (%)
200905211118	Synchronized flow	45	45	45	0.00	0.00
200905211336	Synchronized flow	206	207	206	0.49	0.49
200906301726	Jam flow	104	104	104	0.00	0.00
200907291708	Synchronized flow	59	58	59	1.69	0.00
200907291714	Jam flow	24	24	24	0.00	0.00
200908081012	Synchronized flow	32	32	32	0.00	0.00
200908091114	Synchronized flow	37	37	37	0.00	0.00
200910191706	Jam flow	67	68	68	1.49	1.49
200910221600	Synchronized flow	55	54	55	1.82	0.00
200911120918	Synchronized flow	32	33	32	3.13	0.00
200911191000	Synchronized flow	57	58	58	1.75	1.75
200911191006	Synchronized flow	67	67	67	0.00	0.00
Total/MAPE		785	787	786	0.86	0.31

Table 2 shows the results of scenario II (Bao et al., 2016b). Similar to scenario I, each data set in Table 2 is named by the recording time from Traffic EMS, with a format of year-month-day-hour-minute. Each data set included an hour of continuous traffic flow data. The real hourly traffic volume on the lane located by Traffic EMS was obtained by manual statistics based on continuous video data (video data from 18:00-19:00 and 02:00-03:00 was incomplete; thus the corresponding data sets were missing).

Table 2. Results of experiment scenario II

Data sets	Dominant mode	Counting from Camera (Vehs)	Output of Algorithm 1 (Vehs)	Output of Algorithm 2 (Vehs)	APE of Algorithm 1 (%)	APE of Algorithm 2 (%)
201211240300	Free flow	88	89	85	1.14	3.41
201211240400	Free flow	98	96	97	2.04	1.02
201211240500	Free flow	109	114	112	4.59	2.75
201211240600	Synchronized flow	215	236	233	9.77	8.37
201211240700	Synchronized flow	541	599	583	10.72	7.76
201211240800	Jam flow	663	658	653	0.75	1.51
201211240900	Jam flow	772	723	755	6.35	2.20
201211241000	Jam flow	726	711	727	2.07	0.14
201211241100	Jam flow	743	677	727	8.88	2.15
201211241200	Jam flow	739	651	720	11.91	2.57
201211241300	Jam flow	784	665	773	15.18	1.40
201211241400	Jam flow	713	562	720	21.18	0.98
201211241500	Jam flow	721	632	720	12.34	0.14
201211241600	Jam flow	721	629	735	12.76	1.94
201211241700	Jam flow	642	594	689	7.48	7.32
201211241900	Jam flow	700	625	691	10.71	1.29
201211242000	Synchronized flow	616	632	619	2.60	0.49
201211242100	Synchronized flow	547	563	553	2.93	1.10
201211242200	Synchronized flow	433	463	456	6.93	5.31
201211242300	Synchronized flow	345	357	350	3.48	1.45
201211250000	Free flow	248	245	243	1.21	2.02
201211250100	Free flow	189	188	188	0.53	0.53
Total/MAPE		11467	10838	11557	7.07	2.54

As shown in Table 2, the MAPE of Algorithm 1 for vehicle counting in a continuous 24-h period is 7.07%. By adopting the second adaptive mechanism, Algorithm 2 reduces the MAPE to 2.54%; for any hour period, the accuracy of vehicle counting exceeds 90%. Generally, in free-flow conditions (such as date sets 201211240300, 201211240400, and 201211250100), Algorithm 1 and Algorithm 2 each have a high accuracy of 97%. However, in synchronized flow conditions (such as data sets 201211242100 and 201211242200), the accuracies of Algorithm 1 and Algorithm 2 decrease; however, the results of Algorithm 2 improve. For jam flow, the maximum traffic volume of a single lane with traffic signal control is 600 vehicles; thus, the traffic flow speed was low and the car-following distance between front vehicles and rear vehicles was small. These phenomena caused the “tailgating effect” (Figure 6) in the vehicle waveform data derived from the Traffic EMS. In this case, Algorithm 1 could not eliminate the “tailgating effect” and the accuracy of Algorithm 1 decreased. The APEs of Algorithm 1 increased from 15-20% (refer to data sets 201211241300 and 201211241400). However, Algorithm 2 was capable of solving this type of problem, reducing the counting errors. As shown in Table 2, under jam flow conditions, the hourly APEs of Algorithm 2 were approximately 2%, which satisfied the vehicle counting demands for congested traffic. The experiment results indicate that the proposed vehicle counting methods compensate for the shortage of invasive sensors for vehicle counting.

4. Traffic State Acquisition with Different Layout Strategies of EMS

4.1. Vehicle Speed Estimation with a Single EMS

In this section, the EMS is introduced to obtain the traffic information of vehicle speed. A kind of multifunctional traffic EMS has been designed in our study to collect traffic information such as traffic flow, speed, occupancy, vehicle classification, stopping time, vehicle existence and headway time. This is an advantage to the conventional loop method. The paper proposes a new speed estimation method by a single multifunctional traffic EMS. Compared to traditional methods, this method is simple and has a linear time complexity, so it can be easily calculated on the computer. In practice, this method can obtain a high accuracy.

The multifunctional traffic EMS in this study can collect vehicle speed, vehicle classification, vehicle count, time occupancy, vehicle stopping, vehicle existence and headway time. This section mainly studied the vehicle speed estimation method based on multifunctional traffic EMS and simply discussed the method of vehicle counting, the discrimination of vehicle stopping and existence, and the calculation of stopping time.

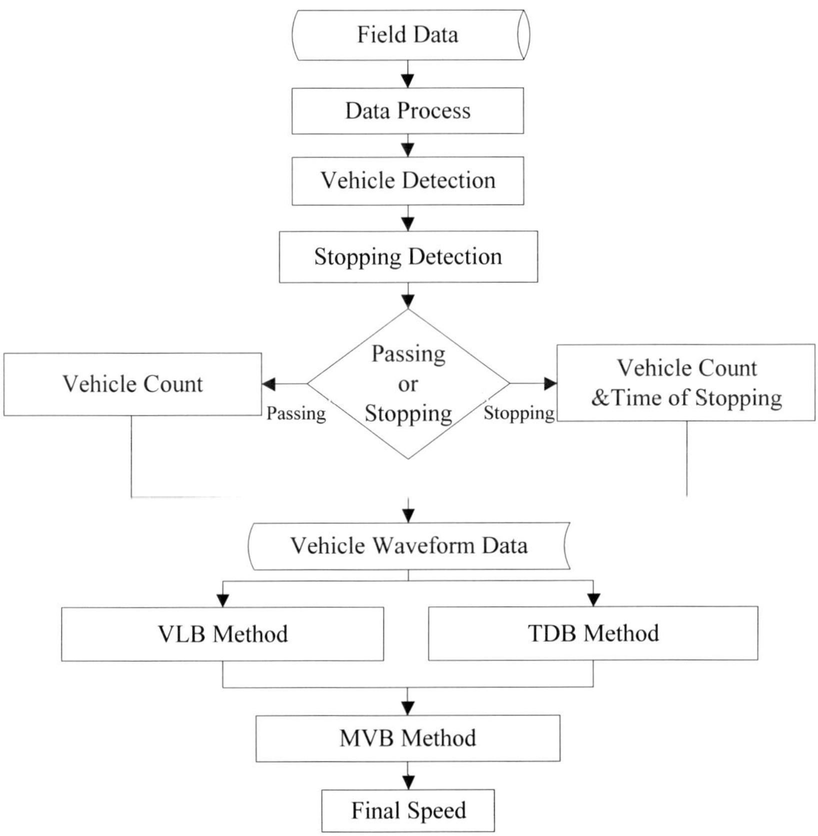

Figure 8. The framework of the speed estimation method.

4.1.1. Framework of Vehicle Speed Estimation

The framework of vehicle detection and speed estimation is shown in Figure 8 (Li et al., 2011). After processing field data, vehicle detection and stopping detection processing was done. Through judging each vehicle passing or stopping, we counted the vehicle and collected vehicle waveform data. Next, the final speed was obtained via estimation methods using Vehicle Length based (VLB), Time Difference based (TDB) and Mean Value based (MVB).

4.1.2. Date Process

The magnetic signal output from the AMR sensors cannot be used directly. It is necessary to design a low-pass filter to eliminate the noise. Therefore, an FIR filter was used for this low-pass design. After the low-pass filter, the data was smoother, better for vehicle detection. To improve the time resolution, the waveform data was processed by linear interpolation. After interpolation, the shape of the vehicle waveform was not changed, while the horizontal axis stretch.

The dual threshold approach was adopted to detect the vehicle. Only when the magnetic amplitude is enough high and holds on for enough long duration, the vehicle is identified. As shown in Figure 5, h_1 and b_1 represents the start amplitude threshold and the start time threshold. At the same time, only when the magnetic amplitude is low enough and holds on a long duration, it is recognized the vehicle is out of range of the detector. In Figure 5, the h_2 and b_2 are used respectively to describe the end amplitude threshold and the end time threshold. area w1 is the vehicle approaching area, and the area w2 is the vehicle leaving area. Those four model parameters h_1, b_1, h_2 *and* b_2 was calibrated by field data. According to the adjusted value of each model parameter, the most suitable value could be obtained through the maximum rate of accuracy of vehicle count. Because those four model parameters are relative values, once they are calibrated, they can be used in other field data.

The process of vehicle detection is described in the following. For no vehicle passing, the magnetic amplitude is the same; for vehicle passing, the magnetic amplitude results in a large change. The point is set the vehicle approaching point P_1 when the absolute difference between magnetic amplitude and yb is greater than h_1 in the waveform from left to right (w1, in Figure 5), and is in range of b_1. By a similar manner, the vehicle leaving point can be found using w2, and the point is written as P_2 (w2, in Figure 5).

How to select point P_1 as the point of vehicle approaching and P_2 as the point of vehicle leaving, the estimation methods of these traffic parameters are described as follows:

For vehicle count: A pair of points (P_1 and P_2) shows that a vehicle has passed the EMS. The waveform between P_1 and P_2 stands for the vehicle data; traffic flow is then increased by 1.

For vehicle stopping: When a vehicle touches a sensor, there will be two states: stopping and passing. The vehicle state can be determined via the shape of the vehicle waveform. If it is the stopping state, the instantaneous speed of the vehicle is set to 0; and if passing state, the instantaneous speed of the vehicle is set to a positive number. When the vehicle state is stopping, the magnetic field around the vehicle has a non-tendency change, but will rise or fall near to some value far from y_b. Figure 9 shows the vehicle waveform of stopping state. A special window (w3) can be set to distinguish if vehicle is stopping. The width and height of the w3 are set to be b_3 and h_3, respectively. If there are successive points in the waveform inside w3, the stopping state exists. To scan the vehicle waveform from left to right using the w3, we find the start-stopping point (SP_1) and end-stopping point (SP_2). The time of stopping can be obtained by the sampling frequency. Each a relative value, b_3 and h_3, can each be calibrated by field data Once calibrated, they can be used in other field data similarly.

For stopping time: The point SP_1 is the start point of vehicle stopping and the point SP_2 is the end point. So the stopping time T_{stop} can be given as:

$$T_{stop}=(SP_2-SP_1)/fs \quad (1)$$

where fs is the sampling frequency; SP_1 is the start sampling point of vehicle stopping; SP_2 is the end sampling point of vehicle stopping.

The state of vehicle existence is very important for traffic security management and military affairs management. Vehicle existence is not different from vehicle stopping. For vehicle existence, when the waveform data are starting to be collected, the vehicle has been there, so vehicle arrival cannot be determined. For vehicle stopping, when the waveform data are starting to be collected, the vehicle has not been there. We know exactly when the vehicle arrives, and we can find the P_1 point and the P_2 point. Using the algorithm of vehicle stopping, we can determine the state of vehicle existence. The vehicle traverses near the sensor, slows down and stops. If the number of P_1 is set to be 1, the number of SP_1 must to be greater than 1. For vehicle existence, if the number of P_1 is set to be 1, the number of SP_1 must be 1, as well. Thus the algorithm can judge the state of vehicle existence and vehicle stopping.

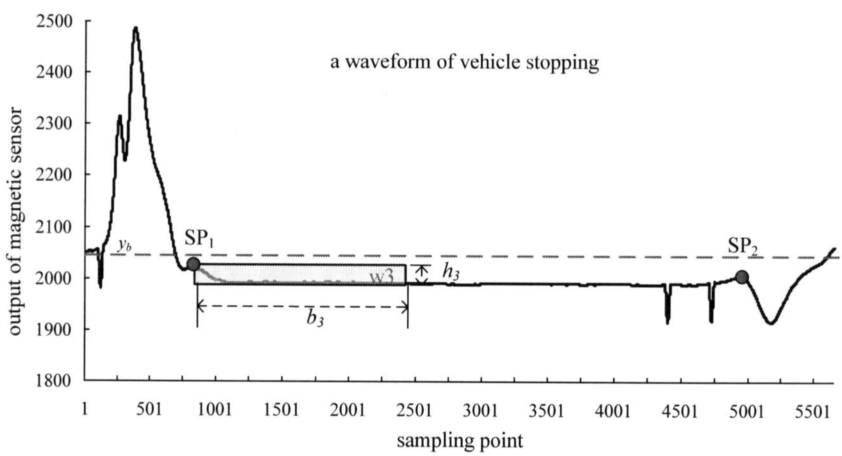

Figure 9. A vehicle waveform of stopping and the distinguish window.

4.1.3. Speed Estimation Methods

4.1.3.1. Vehicle Length Based Speed Estimation

If the length of the vehicle is known, we may calculate the speed according to the time that the vehicle passes the sensor. If the vehicle waveforms of Z1 and Z2 are comparable, using one of the two axes data is ok (in this paper, select Z1). The formula is:

$$v_A = \frac{(L_V + L_e) \times 1000 \times 3.6 \times fs}{P_2 - P_1} \text{(km/h)} \tag{2}$$

where L_V is the length of vehicle; L_e is the extra length of vehicle; fs is the sampling frequency; P_1 is the start sampling point of vehicle detection; P_2 is the end sampling point of vehicle detection.

L_V is calculated by its vehicle classification, and L_V and fs are known parameters. P_2-P_1 can be obtained by vehicle detection algorithm. The parameter L_e needs to be calibrated based on field data.

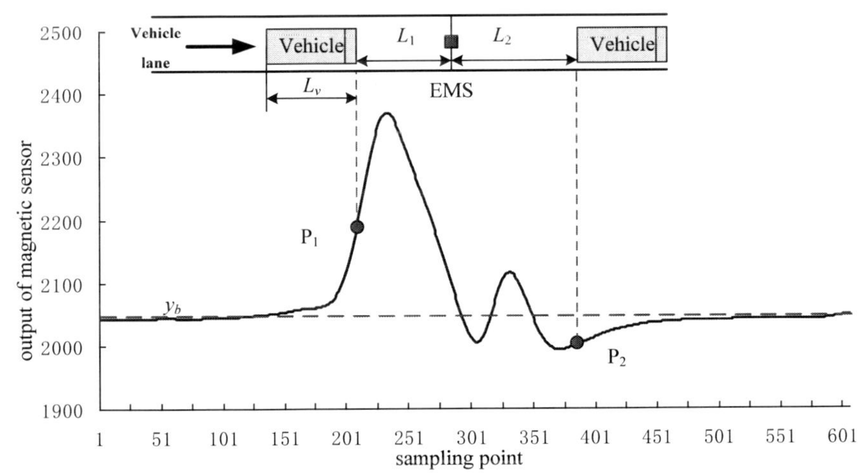

Figure 10. Diagram of extra vehicle length.

Because the magnetic amplitude at P_1 is not equal to the magnetic amplitude when the vehicle head is passing over the sensor and the magnetic amplitude at P_2 is also not equal to the magnetic amplitude when the vehicle tail is just passing over the sensor, thus the extra length L_e is existent. Figure 10 shows the existence of L_e. When the magnetic amplitude is equal to the magnetic amplitude at P_1, the distance between the vehicle and sensor is L_1. When the magnetic amplitude is equal to the magnetic amplitude at P_2, the distance between vehicle and sensor is L_2. Then $L_e = L_1 + L_2$. For different thresholds, the L_e will be different.

According to the characteristic of field magnetic, we know that the waveform of the same vehicle is invariant under different speeds, while the time axis is compressed or stretched. And for the same speed, the waveforms of the different vehicles were found to vary. So vehicles with a same vehicle type must also have the same L_e, and vehicles with a different type may have a different L_e. Based on the field data of different vehicle types, the L_e of different vehicle types can be obtained by the real speed. In practice, the L_e of different type vehicle is close, so we can use an average L_e of different type vehicle as a common L_e.

4.1.3.2. Time Difference Based Speed Estimation

For the EMS with two chips, when a vehicle is passing these two chips (Z1 and Z2), two similar waveforms will be collected. Derived from the distance between these two chips and from the time difference of the two similar waveforms, the speed of the vehicle can be obtained. Because the distance between the two chips is not large enough for accuracy speed estimation, the time difference will be very small. In order to reduce the round error, we can do some processing of the waveform data. The effective method to deal with the data is through an interpolation method. the formula is given as follows:

$$v_B = \frac{L \times fs \times 3.6 \times k}{ds} \quad \text{(km/h)} \tag{3}$$

where L is the distance between the two chips; fs is the sampling frequency; k is the times of interpolation; and ds is the sampling difference of the two similar waveforms.

After interpolation, the sampling difference (ds) of the two similar waveforms is shown in Figure 11 (the times of interpolation is 10) and can be obtained as following methods.

Method 1: Moving the abscissa of the two waveforms, we can write down the abscissa difference and calculate the correlation coefficient of the two data vectors. We will select the abscissa difference as the sampling difference (ds) when the value of the correlation coefficient is at most one.

Method 2: Take some special points in the two waveforms, and calculate corresponding abscissa difference of those points. The sampling difference (ds) is the average of these abscissa differences.

The quotient is the sampling difference divided by the times of interpolation and the sampling frequency is the time differential of the vehicle passing each of the two chip.

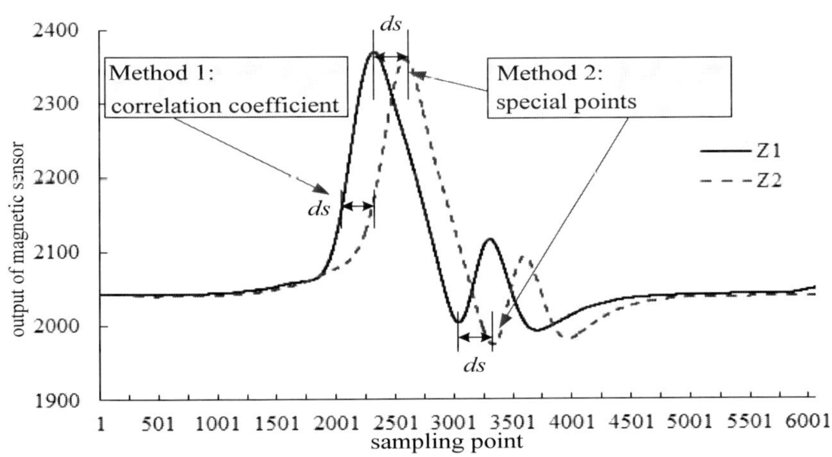

Figure 11. Parameter *ds* of two similar waveforms with same vehicle (Z1 axis and Z2 axis).

4.1.3.3. Mean Value Based Speed Estimation

Mean value always possesses some good characters, and it can eliminate some errors. So if the speeds of v_A (VLB) and v_B (TDB) are collected, the Mean Value based speed v_C can be given as follows:

$$v_C = \frac{v_A + v_B}{2} \text{ (km/h)} \tag{4}$$

4.1.4. Experiments

4.1.4.1. Data Collection and Process

The experiment selects 45 vehicles' waveform data collected at 11 o'clock on May 12, 2009, in Beijing. We examined the traffic situation using a live video, and obtained the vehicles' state and speed when it passes a sensor. Starting time is set to 0, so the relative time when a vehicle passes over the sensor can be easily obtained. The average length of each vehicles set to 4.5m. The FIR low-pass filter is designed to process the field data.

4.1.4.2. Detection of Vehicle Approaching and Leaving

By setting the suitable size of the approaching window (h_1 and b_1) and the leaving window (h_2 and b_2), the waveform data was cut out from the data of all vehicles. The parameters of vehicle detection w1 and w2 were set as follows: h_1= 70, b_1 = 30, h_2 = 25, b_2= 120. And the parameters of stopping identification window was set as followed: h_3 = 10, b_3 = $MinStopT \times fs$. *MinStopT* was set as 2s (the minimum stopping time). Using these window parameters, we collect all 45 vehicles' waveforms including vehicle waveforms with different states. These waveforms reflect the state when the vehicles passes over a sensor.

4.1.4.3. Speed Estimation and Analysis

Based on the vehicle waveform data and the speed estimate methods, three reference speeds using different estimation methods were calculated. According to the calibrated field data, the common L_e was found to be

1.8m. Table 3 shows the result of different speed estimation methods. The MAE of VLB, TDB and MVB is, respectively, 4.12km/h, 5.90km/h and 4.05km/h. Moreover, the accuracy rate of vehicle counts nearly 100% and the number of vehicle stopping is 3 times.

Table 3. Real speed and estimation speed of VLB, TDB and MVB (km/h)

Vehicle Number	Real Speed	v_A	v_B	v_C	v_A-MAE	v_B-MAE	v_C-MAE
1	33	31	20	25	1.6	12.6	7.1
2	47	34	33	34	12.9	13.6	13.2
3	19	13	22	18	6.2	3.0	1.6
4	0 (Stopping)	0 (Stopping)	0 (Stopping)	0 (Stopping)	0.0	0.0	0.0
5	8	8	11	10	0.1	3.2	1.7
6	13	30	29	29	16.8	15.4	16.1
7	14	12	14	13	1.9	0.4	0.8
8	34	33	25	29	1.1	9.1	5.1
9	31	29	22	26	2.2	9.1	5.6
10	25	30	18	24	5.0	6.8	0.9
11	30	44	40	42	14.0	9.9	12.0
12	27	37	17	27	10.2	10.1	0.0
13	11	10	18	14	0.9	7.3	3.2
14	29	21	25	23	7.8	3.9	5.9
15	0	0	0	0	0.0	0.0	0.0
16	12	14	8	11	2.3	4.0	0.9
17	19	17	22	20	1.7	3.4	0.8
18	17	13	18	16	3.7	1.5	1.1
19	27	40	22	31	13.2	4.6	4.3
20	30	25	22	24	5.0	7.8	6.4
21	34	27	25	26	7.1	9.1	8.1
22	19	16	20	18	2.7	1.2	0.8
23	26	26	29	27	0.1	2.7	1.4
24	22	16	25	20	6.1	2.9	1.6
25	21	17	20	18	3.8	0.9	2.3
26	31	32	22	27	0.8	9.1	4.1
27	31	33	25	29	1.8	6.3	2.3

Table 3. (Continued)

Vehicle Number	Real Speed	v_A	v_B	v_C	v_A-MAE	v_B-MAE	v_C-MAE
28	14	10	22	16	3.9	8.3	2.2
29	0	0	0	0	0.0	0.0	0.0
30	13	19	17	18	5.6	3.2	4.4
31	14	18	15	17	3.9	1.2	2.5
32	16	16	29	22	0.4	12.9	6.6
33	20	15	17	16	4.7	3.1	3.9
34	27	21	25	23	5.8	1.8	3.8
35	24	29	33	31	4.8	9.1	7.0
36	33	28	29	28	4.6	4.1	4.3
37	33	34	25	29	1.4	7.6	3.1
38	23	16	20	18	6.7	2.8	4.7
39	11	10	11	11	0.9	0.2	0.3
40	17	19	22	21	1.6	4.7	3.2
41	23	21	22	22	1.7	0.5	1.1
42	28	27	20	23	0.8	7.8	4.3
43	47	42	29	35	4.9	18.3	11.6
44	9	12	29	20	3.2	19.7	11.4
45	5	3	7	5	1.5	2.1	0.3
Average	21.40	21.07	20.49	20.78	4.12	5.90	4.05

There are some v_A-MAEs greater than 10km/h such as vehicles 2, 6, 11, 12, 19. Some factors may influence the accuracy rate of the VLB method. The length of each vehicle was assumed to be 4.5m for short, but may not be representative of each vehicle. If the length of each vehicle could be obtained exactly and easily through vehicle classification or some other methods, the accuracy rate of VLB would greatly increase. For the TDB method, there are also some v_B-MAEs greater than 10km/h such as vehicles 1, 2, 6, 12, 32, 43, 44. Factors, such as the times of interpolation, interpolation algorithm, noise of waveform data, and the distance between the two chips, influence the accuracy rate of TDB. Even more, the EMS of our experiment was installed near the traffic light, which increases the ability to completely remove the influence of the magnetic field for each vehicle. If the field data was collected from the highway or by a far from the traffic light, the accuracy would be better. Besides, the MVB method is often better than other methods (the MAE 4.05 is less than both of 4.12 and 5.90), because it can always reduce the maximum error of the other two methods. What should be pointed out is that only when the estimation speeds of the other two methods are good enough, the MVB method is meaningful and effective. This is because sometimes one of VLB and TDB will be better than MVB. While the real average speed of the 45 vehicles is 21.04 km/h, the average speed of v_A is 21.07 km/h and v_C is 20.78 km/h as shown in Table 3.

4.2. Vehicle Classification with Networked EMS

The main objective of this section is to develop a more effective vehicle classification method using a single multi-functional EMS for an urban road. In this paper, a novel method is proposed to classify vehicles by measuring and processing field magnetic signals based on a single multi-functional sensor we designed and developed. We first proposed the on-line vehicle classification method with a single multi-functional EMS and optimal MNS-based CART. The CART-based model had a simple structure and fast execution efficiency. And the multi-functional EMS was superior to the existing sensors in terms of reliability, cost, size and function. Furthermore, by comparison with current methods, the proposed method was competitive in terms of accuracy rate, sample robustness and execution time. Finally, experiments based on vehicle waveform data in a field road were implemented to test the performance of the new method (Li et al., 2014a).

4.2.1. Data Acquisition

The wireless EMS, which was designed and developed by our group, was used to acquire the vehicle waveform data. Its shape parameters are diameter D = 100mm and height H = 64mm. The sensor is capable of detecting real-time magnetic field strength. Making use of the magnetoresistance effect, it translates the magnetic field strength into mV level voltage signal, and then into V level voltage signal by an amplifier. A filter is used to reduce environment noise. Then a digital output of the sensor was obtained by an A/D converter. The values of the digital output correspond to the variations that a vehicle disturbs the magnetic field. If no vehicle is nearby a sensor, the digital output of the sensor would read an invariable value (eliminating the stochastic noises), which is defined as the benchmark value y_b. When a vehicle passes the sensor, for the effect of ferromagnetic objects to magnetic field, the outputs of the sensor reads superposition values of the benchmark value y_b and the variations caused by the vehicle, is defined as the vehicle's waveform. The multi-functional sensor can detect several traffic flow parameters such as vehicle existence, vehicle stop, vehicle retrogradation, vehicle classification, traffic count, vehicle speed, and occupancy ratios.

Due to the influences of the environment and the sensor itself, the vehicle waveform data contains some high frequency noises (the sampling frequency f_s=333Hz for vehicle classification data). Therefore, a low-pass filter should be designed to eliminate high frequency noises. By analyzing the frequency distribution of noises, we designed a low-pass filter to obtain smooth vehicle waveforms.

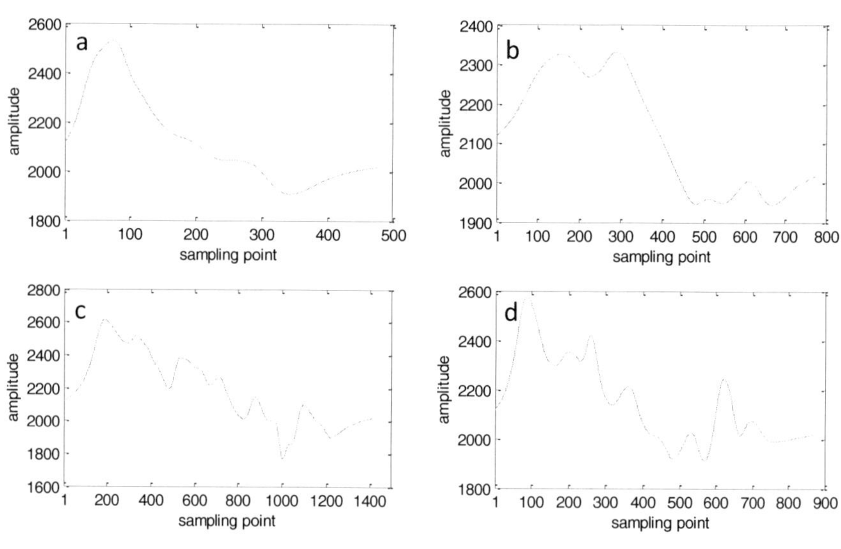

Figure 12. Different vehicle waveforms (a and b, cars; c and d, busses).

With a low-pass filter, the vehicle waveform was smoother and better for vehicle detection and feature extraction. A dual-window vehicle detection algorithm was used to extract the features of each vehicle waveform.

The integral waveform of each vehicle was obtained using the dual-window vehicle detection algorithm. Figure 12 shows some integral vehicle waveforms. For the differences of spatial distribution of ferromagnetic objects, the waveform features of each vehicle was different, which is the foundation for vehicle classification.

4.2.2. Feature Extraction

A vehicle waveform contains several characteristics. The characteristics that are easy to compute and are useful for vehicle classification was extracted. When the same vehicle passes the sensor at different speeds, the vehicle's waveforms acquired at these times would not be identical under a fixed sampling frequency. Since the time interval that a vehicle passes a sensor is small, the change in vehicle speed may sometimes be ignored. Therefore, it can be concluded that a vehicle passes a sensor at a constant speed each time.

Then, for the same vehicle, the differences in the vehicle's waveforms acquired at different times will be stretched or compressed at the time axis. In order to eliminate the effect of speed on vehicle classification, some speed-independent features will be introduced. In this section, 8 time-domain features were selected as the model inputs.

Those features and their notations are: the number of peaks (x_1), the maximum peak time ratio (x_2), the minimum trough time ratio (x_3), the mean value (x_4), the standard deviation (x_5), the maximum peak amplitude (x_6), the minimum trough amplitude (x_7) and the maximum peak/trough amplitude ratio (x_8). Denote vector $\mathbf{s}$ as a vehicle waveform series, that is $\mathbf{s}=[s_1,s_2,...,s_n]$, where $s_1,s_2,...,s_n$ is the sampling point of a vehicle waveform, and n is the number of the sampling points.

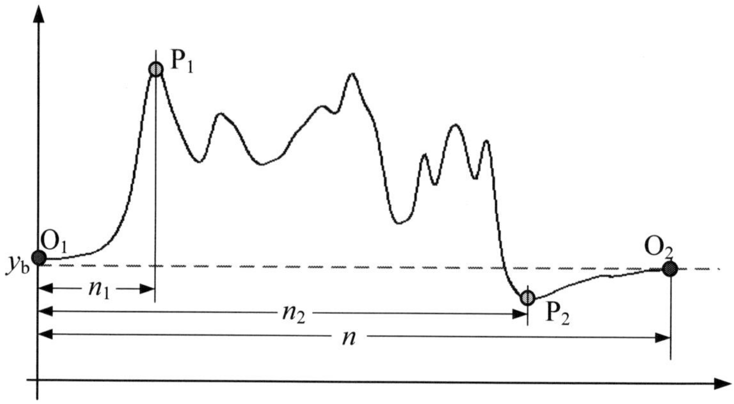

Figure 13. A vehicle waveform and its time-domain features.

Figure 13 shows the waveform of a vehicle and its time-domain features. O_1, O_2 are the start point and the end point, respectively, of the vehicle waveform that were obtained by Algorithm 1; P_1, P_2 are the maximum peak point and minimum trough point; n_1, respectively, n_2 is the number of sampling points from O_1 to P_1 and O_1 to P_2, respectively.

Denote n_p, n_t as the number of peaks and troughs, respectively, we have $|n_p\text{-}n_t|=0$ or 1. Then: $x_1=n_p$, $x_2=n_1/n$, $x_3=n_2/n$, x_4=ave(**s**), x_5=var(**s**), x_6=max(**s**), x_7=min(**s**), x_8=max(**s**)/min(**s**), where ave, var, max, min are the mean function, the standard deviation function, the maximum function and the minimum function of vector **s**, respectively.

Denote **x** as the feature vector of a vehicle waveform, then $\mathbf{x}=[x_1,x_2,x_3,x_4,x_5,x_6,x_7,x_8]$. Features such as x_1, x_2, x_3, x_6, x_7, x_8 reflect the concentration and relative position of ferromagnetic material of a vehicle, the inherent property of the vehicle and therefore they are speed-independent. Features such as x_4, x_5 reflect the average distribution density and uniformity of ferromagnetic material. While different vehicle speeds lead to different numbers of sampling points, the average distribution density and uniformity of ferromagnetic material will remain constant. The sensor outputs are uniform sampling based on both time and space. Therefore, the features x_4 and x_5 are speed-independent.

Each vehicle waveform was replaced by 8 speed-independent features. Compared to traditional methods, these features eliminate some errors caused by different vehicle speeds. When these 8 speed-independent features are selected as the model inputs, the time and space complexities of algorithm will be greatly reduced. So then, the algorithms of vehicle classification will be implemented easily in sensor nodes with limited memory and energy sources.

4.2.3. Methodology

Decision tree is a very efficient machine learning classification model and it uses a binary tree to represent processing logic. The basic principle of a decision tree is that it recursively splits the data into subsets so that each subset contains a similar state or region of the data features. When this recursive process is done, a decision tree is created. Creating a decision tree is a process of tree growth that ceaselessly splits data sets. Each split corresponds to a decision problem, in addition to a node. For each split, a Max 'difference' between a subset and the rest of data set is required.

Algorithm 3: the CART algorithm	
1:	Create a root node R of sample set T

2:	**if** All the samples of T are the same class or T only contains τ samples
	then
	The node R is a leaf node, and return
	else
	Go to step 3
	end if

3:	For each feature x_i of T ($1<=i<=N$, N is the number of features, in this paper N=8), compute GINI values of different split cases, and denote the minimum GINI of x_i as GINI(i)
4:	**for** $i=1 \rightarrow k$ Find the minimum GINI(i), select x_i as the decision feature of node R, and record the split case **end for**
5:	According to the split case in step 4, split sample set T into two subset T1 and T2
6:	Repeat step 1 to 5 for T1
7:	Repeat step 1 to 5 for T2
8:	A decision tree has been created

A decision tree is composed of decision nodes, branches and leaves. The top node of decision nodes is the root node and each branch generates a new decision node or leaf. Each decision node represents a problem or decision, and generally corresponds to a feature of an object in order to be classified. Each leaf node represents a possible classification result.

In this paper, the optimal MNS-based CART algorithm is used to create a decision tree. The CART algorithm adopts a binary recursive partitioning technology to split current sample set into two subsets, which makes the resulting decision tree have two branches for each non-leaf node. The decision tree which is based on CART algorithm, is a binary tree with a simple structure. The CART algorithm is provided in Algorithm 3.

The algorithm parameter τ is the MNS which relates to the fitting degree of training samples based decision tree model. When τ = 1, it will be the optimal division for the training samples, but it will result in an overfitting decision tree. An overfitting decision tree is not appropriate to classify new samples and will lead to low accuracy rates. Therefore, it is necessary to find an optimal τ for CART algorithm. Fortunately, the sample set is always limited and the model learning is an offline process. We can find the optimal τ by computing the classification accuracy rates of all samples based on different decision trees trained with different values of τ. The value of τ corresponding to the maximum accuracy rate will be the optimal one.

Another algorithm parameter GINI is a measure of impurity for a sample set or subset: GINI of sample set T is

$$\mathrm{GINI}_{\mathrm{T}} = 1 - \sum_{k=1}^{c} p(k)^2 \tag{5}$$

where k represents the kth class in set T, $1 \leq k \leq c$; c is the number of total classes; $p(k)$ is the (possibly weighted) relative frequency of class k in set T, and $\sum_{k=1}^{c} p(k) = 1$.

Assume that a split s of some feature divides set T into two subsets T1 and T2, and then the new GINI generated by s will be

$$\mathrm{GINI}_{\mathrm{T}}(s) = q \times \mathrm{GINI}_{\mathrm{T1}} + (1-q) \times \mathrm{GINI}_{\mathrm{T2}} \quad (6)$$

where q is the (possibly weighted) fraction of samples going T1.

CART can achieve self-learning according to the input features, and can select the most effective features for growing a decision tree. CART uses a prior's mechanism. Priors are akin to target class weights but are invisible in that they do not affect any counts reported by CART in the tree. Rather priors automatically reweights the data to balance the classes, and ensures that the tree selected as optimal minimizes balanced class error. The reweighting is implicit in the calculation of all probabilities and improvements and requires no user intervention [17]. To provide this flexibility, CART is unnecessary to those steps of eliminating invalid features and calculating feature correlation which are essential steps in other classification algorithms such as Artificial Neural Network or Naive Bayes.

An optimal decision tree (T_o) which is able to effectively classify the known samples can be established by training those samples. Then the online classification process can be executed by making use of the T_o. Denoting m as the number of training samples, we can obtain the feature matrix which is denoted as $\mathbf{X}_{m\times 8}$, as is $\mathbf{X}_{m\times 8}=[\mathbf{x}_1,\mathbf{x}_2,\ldots,\mathbf{x}_i,\ldots,\mathbf{x}_m]^{\mathrm{T}}$. So$\mathbf{x}_i$ is the feature vector of ith vehicle waveform and $\mathbf{x}_i=[x_{i1},x_{i2},x_{i3},x_{i4},x_{i5},x_{i6},x_{i7},x_{i8}]$, $i=1,2,\ldots,m.,y_i$ is the corresponding vehicle class of ith vehicle waveform and the vehicle class vector is e $\mathbf{y}=[y_1,y_2,\ldots,y_i,\ldots,y_m]^{\mathrm{T}}$.

By training sample feature matrix $X_{m\times 8}$ and its corresponding vehicle class vector $\mathbf{y}$ with different values of τ, a series of decision trees can be created. The decision tree which has a maximum classification accuracy rate of all samples will be selected as the T_o, while the corresponding value of τ is the optimal τ. The structure of a decision tree is A triangle represented by a decision node with having two branches and a little dot that represents a leaf node, the vehicle class.

Due to possible over-fitting for T_o, it will be necessary to create an optimal pruning tree PT_o by pruning a T_o on occasion. The PT_o will then be more robust to new samples. CART algorithm uses afterwards pruning method. By pruning branches on an over-fitting tree, the unpruned nodes at the bottom will be denoted as leaf nodes. This paper adopts a MEP-based rule to prune T_o and selects a pruning tree with a maximum classification accuracy rate of the test samples as the optimal pruning tree PT_o.

Making use of the T_o and PT_o, it will be easy to implement a vehicle classification algorithm in a sensor node and to achieve on-line classification for urban traffic. In the trees of T_o and PT_o, the optimal features can be revealed. In order to reduce computation, calculating those optimal features will be enough for on-line vehicle classification in practice. On-line vehicle classification is an ergodic process along the decision tree from top to bottom. In the process, it will jump to the next branch based on the decision result of each decision node. The process will eventually reach a leaf node. Reading the information of the leaf node, a class of the new sample can be obtained.

4.2.4. Experiments

4.2.4.1. Vehicle Samples

The EMS was laid in the center of a lane of Jiadadonglu road, Haidian district to collect vehicle waveform data. From May 2009 to November 2009, we collected vehicle waveform data of 470 cars and 72 busses, and used the video to record the corresponding vehicle classes. These 542 vehicle waveforms were selected as sample sets to test our method.

After acquiring vehicle waveform data and extracting the feature extraction we obtained a feature matrix of each of the 542 vehicle waveforms (denoted as $\mathbf{X}_{542\times 8}$). In order to train and test different methods, 271 vehicle waveform (including 235 cars and 36 busses) samples were selected randomly as the training samples in this paper (denoted as $\mathbf{X}_{train}$). The remaining samples were test samples (denoted as $\mathbf{X}_{test}$). Concurrently, we used the videos to count the actual vehicle classes. Denoted with 1 as the car, 2 as the bus, and we then found the vehicle class vector ytrain that corresponded to $\mathbf{X}_{train}$ and the vehicle class vector y_{test} that corresponded to $\mathbf{X}_{test}$.

4.2.4.2. Optimal MNS τ

In order to obtain the optimal τ, accuracy rates of samples $\mathbf{X}_{train}$ and $\mathbf{X}_{test}$, which are based on different decision trees with differing t t values, were computed (Table 4).

Table 4. Accuracy rates of samples $\mathbf{X}_{train}$ and $\mathbf{X}_{test}$ based on different decision trees with different τ (Li et al., 2014a)

τ	Accuracy rate of $\mathbf{X}_{train}$			Accuracy rate of $\mathbf{X}_{test}$		
	bus	car	average	bus	car	average
1	1.000	1.000	1.000	0.750	0.987	0.869
2	1.000	1.000	1.000	0.750	0.987	0.869
3	1.000	1.000	1.000	0.750	0.987	0.869
4	1.000	0.996	0.998	0.778	0.987	0.883
5	1.000	0.996	0.998	0.778	0.987	0.883
6	1.000	0.996	0.998	0.778	0.987	0.883
7	1.000	0.996	0.998	0.778	0.987	0.883
8	1.000	0.996	0.998	0.778	0.987	0.883
9	1.000	0.996	0.998	0.778	0.987	0.883
10	0.972	0.996	0.984	0.778	0.996	0.887
11	0.944	0.996	0.970	0.778	1.000	0.889
12	0.944	0.996	0.970	0.778	1.000	0.889
13	0.944	0.996	0.970	0.778	1.000	0.889
14	0.944	0.996	0.970	0.778	1.000	0.889
15	0.944	0.996	0.970	0.778	1.000	0.889
16	0.944	0.996	0.970	0.778	1.000	0.889
17	0.944	0.996	0.970	0.778	1.000	0.889
18	0.944	0.996	0.970	0.778	1.000	0.889
19	0.944	0.996	0.970	0.778	1.000	0.889
20	0.944	0.996	0.970	0.778	1.000	0.889
36*	0.944	0.991	0.968	0.778	1.000	0.889
235**	0.944	0.991	0.968	0.778	1.000	0.889
271***	0.917	0.991	0.954	0.778	1.000	0.889
> 271	0.000	1.000	0.500	0.000	1.000	0.500

Note: *number of busses in $\mathbf{X}_{train}$, **number of cars in $\mathbf{X}_{train}$, ***number of cars and busses in $\mathbf{X}_{train}$.

As shown in Table 4, when τ is lower, the classification accuracy rate of $\mathbf{X}_{train}$ will be 100%, while the accuracy rate and adaptability of $\mathbf{X}_{test}$ is poor. As τ increases, the accuracy rate of $\mathbf{X}_{train}$ falls; however, the accuracy rate of $\mathbf{X}_{test}$ can be improved. When τ is high enough, the decision tree will not be able to classify the samples. In principle, τ should not be higher than the fewest number of samples. Otherwise, it will make against the classification of a class of samples with fewest number. In this paper, the averages of the two accuracy rates, $\mathbf{X}_{train}$ and $\mathbf{X}_{test}$, will be computed using every possible τ *value*; the τ with the highest average rate will be the optimal one.

By computing the average classification accuracy rates of sample set $\mathbf{X}_{542\times 8}$, based on decision trees with differing values of τ, it can be drawn that when τ is between 4 and 9, we find the maximum average accuracy rate. The greater the value of τ is, the faster the speed of algorithm converge will be; therefore, the optimal τ is 9 in this paper.

4.2.4.3. Results

When τ=9, through training the samples of $\mathbf{X}_{train}$ and $\mathbf{y}_{train}$, the optimal decision tree T_o can be obtained based on CART. In Figure 14, the triangles represent decision nodes, the dots represent leaf nodes, and 1 represents the result of vehicle classification is a car and 2 a bus.

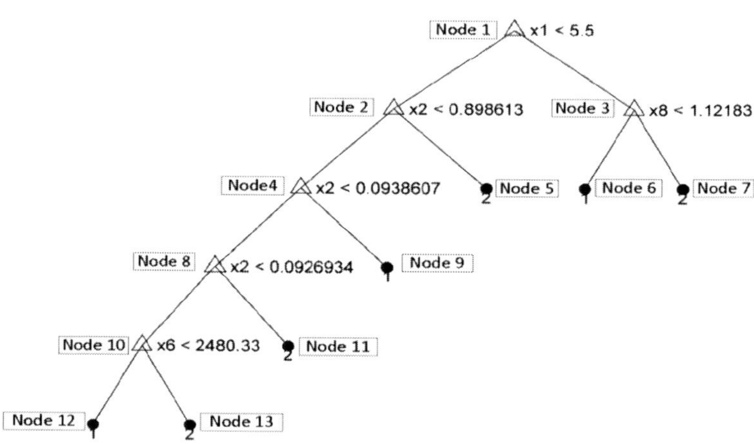

Figure 14. Optimal decision tree based on CART by training the training samples.

From tree T_o, we have the best combination of waveform features x_{opt}=[x1,x2,x6,x8]. Using the decision tree T_o to implement on-line vehicle classification, we calculated the four features x1, x2, x6 and x8. Derived from Figure 14, the on-line classification algorithm (Algorithm 4) is shown as follows.

By using Algorithm 4, we respectively take the sets of training samples $\mathbf{X}_{train}$ and test samples $\mathbf{X}_{test}$ to test the validity and effectiveness of our method. The accuracy rates of $\mathbf{X}_{train}$ and $\mathbf{X}_{test}$ are shown in Table 5 and Table 6.

Algorithm 4: the on-line classification algorithm		
1:	node 1	if x1<5.5 then node 2 else node 3
2:	node 2	if x2<0.898613 then node 4 else node 5
3:	node 3	if x8<1.12183 then node 6 else node 7
4:	node 4	if x2<0.0938607 then node 8 else node 9
5:	node 5	class = 2
6:	node 6	class = 1
7:	node 7	class = 2
8:	node 8	if x2<0.0926934 then node 10 else node 11
9:	node 9	class = 1
10:	node 10	if x6<2480.33 then node 12 else node 13
11:	node 11	class = 2
12:	node 12	class = 1
13:	node 13	class = 2

Table 5. Results of training samples $\mathbf{X}_{train}$ based on T_o

	Cars (Tree output)	Busses (Tree output)	Accuracy rate / %
Cars (235)	234	1	99.6
Busses (36)	0	36	100.0
Average			99.8

Table 6. Results of test samples $\mathbf{X}_{test}$ based on T_o

	Cars (Tree output)	Busses (Tree output)	Accuracy rate / %
Cars (235)	232	3	98.7
Busses (36)	8	28	77.8
Average			88.3

In order to improve the classification accuracy of the test samples, we obtained the optimal pruning tree PT_o by pruning the T_o based on the MEP rule. As shown in Figure 15, the dotted lines contains some parts of the T_o and the remaining lines are the PT_o.

Based on PT_o in Figure 15, a pruning tree based on-line classification algorithm is shown in Algorithm 5.

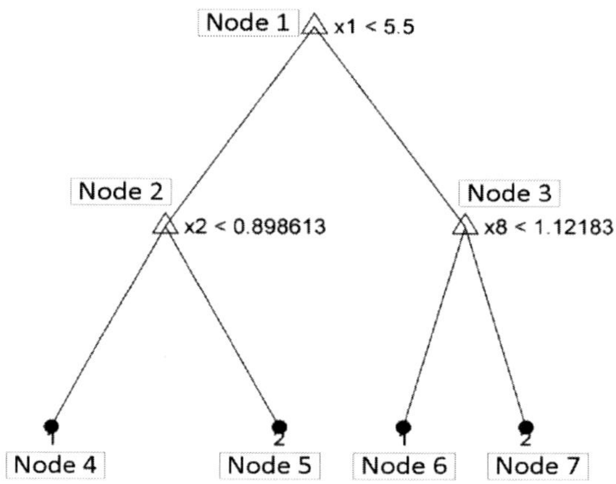

Figure 15. Optimal pruning tree PT_o.

Algorithm 5: the pruning tree based on-line classification algorithm		
1:	node 1	if x1<5.5 then node 2 else node 3
2:	node 2	if x2<0.898613 then node 4 else node 5
3:	node 3	if x8<1.12183 then node 6 else node 7
4:	node 4	class = 1
5:	node 5	class = 2
6:	node 6	class = 1
7:	node 7	class = 2

Table 7. Results of training samples X_{train} based on PT_o

	Cars (Tree output)	Busses (Tree output)	Accuracy rate / %
Cars (235)	234	1	99.6
Busses (36)	2	34	94.4
Average			97.0

Table 8. Results of test samples $\mathbf{X}_{test}$ based on PT_o

	Cars (Tree output)	Busses (Tree output)	Accuracy rate / %
Cars (235)	235	0	100.0
Busses (36)	8	28	77.8
Average			88.9

Compared to Algorithm 4, Algorithm 5 is more concise and with a lower computational complexity, which is very important in implementing on-line vehicle classification in a sensor node with very limited arithmetic and memory capacity. To adopt the MEP-based rule to execute pruning operation of To, Algorithm 5 was found to have a higher classification accuracy and robustness for new samples than Algorithm 4. Based on Algorithm 5, test results of $\mathbf{X}_{train}$ and $\mathbf{X}_{test}$ are shown in Table 7 and Table 8, respectively. As we can see, the accuracy rate of $\mathbf{X}_{train}$ that is based on PT_o has seen a slight decline, while the accuracy rate of $\mathbf{X}_{test}$ was improved. Improving the accuracy of new samples is quite important to on-line vehicle classification on the field roads.

4.3. Vehicle Queue Length Estimation at Signalized Intersections with EMS

Magnetic sensors have been used to obtain traffic information through quantitative and qualitative analysis of the magnetic field. Traffic information includes vehicle detection (He et al., 2014; Yang and Lei, 2015), vehicle counting (Kwong et al., 2010; Taghvaeeyan and Rajamani, 2014), travel time (Kwong et al., 2010; Sanchez et al., 2011), vehicle speed estimation (Li et al., 2011; Taghvaeeyan and Rajamani, 2014), vehicle classification (He et al., 2014; Li et al., 2014b; Taghvaeeyan and Rajamani, 2014; Yang and Lei, 2015), and even the strain state of vehicle structure (Gontarz et al., 2015). Also, magnetic sensors can acquire the possibility of collision detection (Gontarz et al., 2015; Taghvaeeyan and Rajamani, 2014). This section focuses on queue length estimation methods by considering different layout strategies of magnetic sensors and lane features at signalized intersections.

Queue length is the basic data used in signal cycle optimization, traffic capacity control and traffic LOS evaluation at signalized intersections. Meanwhile, it is also an important input parameter for traffic signal control systems. As the development of ITS in modern cities is rising all over the world, the large-scale traffic information perception (including queue length estimation) in urban road network will be necessary and urgent. With a magnetic sensor deployed in the center of a lane, many traffic parameters (vehicle approach and departure time, traffic volume, vehicle speed, etc.) can be obtained (H Li et al., 2014b; Li et al., 2011). When one direction of a one-way entrance has more than one lane, the sensors in the same cross-section are treated as a single cross-section sensor. The work of low-cost traffic information perception methods will be meaningful and worthy. The objective of this study is to estimate queue length by considering both the economy and availability of magnetic sensors. The queue length perception mechanisms using different layout strategies of magnetic sensors will be modelled and revealed. More attention will be paid to the single sensor-based strategy for its efficient detection and low cost. After a single cycle, the queuing status of the vehicles can be estimated using the velocities and interval times of these vehicles. Because of one single sensor, the single sensor-based strategy has advantages of being relatively inexpensive, requires a quick installment, has a concise wireless network structure, and is convenient in collecting field data for large-scale networks and intersections.

Based on the number of exit directions of a one-way entrance at a signalized intersection, the entrance of an intersection can be categorized by three types: one-direction, two-direction, and three-direction. Straight-left lane and straight-right lane is treated as straight lane because they share the same control mechanism. Turning lanes typically share the same lane with left-turn, so it won't be considered individually. One-direction type is common in small-scale intersections, signalized expressways or freeway ramps. Two-direction type is common in T-intersections or intersections with right- or left-turn prohibition. Estimating queue length for two-direction type intersections can be sorted into three different cases: two-direction controlled by a same signal phase, two-direction controlled separately, and one-direction controlled only. Three-direction type is common in ordinary intersections (Each direction has one lane) or large intersections (Each direction has more than one lane). The control mechanisms of the three-direction type typically fall under four cases:

Case 1: the left-turn and through lanes controlled by a same signal phase with an uncontrolled right-turn lane (Common in two-phase signalized intersections);

Case 2: the left-turn and through lanes controlled separately with an uncontrolled right-turn lane (Common in four-phase signalized intersections);

Case 3: all of the three directions controlled by a same signal phase (Common in ordinary two-phase signalized intersections with right-turn controlled);

Case 4: the right-turn and through lanes controlled by a same signal phase with separately controlled right-turn lane (Common in four-phase signalized intersections with right-turn controlled).

This section presents appropriate sensor layout strategies by considering the practical detection precision demands and the queue length estimation methods (H Li et al., 2017). Some methods are used for real-time queue length estimation, vehicle state identification, signal optimization, and LOS evaluation at intersections; while others are used for queue length estimation of the immediate past signal cycle and basic traffic data acquisition.

4.3.1. Layout Strategy 1: Uniform Deployment

Layout strategy 1 deploys sensors at a distance interval of $\triangle l$; the distance between the first sensor and the stop line is l_0 (Figure 16). If the total number of the deployed sensors equals n, then the length between stop line and the nth sensor is $L=l_0+(n-1)\Delta l$. This section first considers the least complex condition with one-direction, as shown in Figure 16(a). The sensors are numbered in the opposing direction of vehicle flow. If the kth sensor detects a stopped vehicle and if a sensor with a higher number has not detected stopped vehicles, then the relationship between queue length estimation value l_{qest} and k is shown as Eq. (7).

$$l_{qest}(k) = l_0 + (k-1)\Delta l \tag{7}$$

The maximum absolute error between l_{qest} and the real queue length l_q is $\Delta l/2$, and the relative error is $\Delta l/(2l_q)$. If the nth sensor detects a stopped vehicle, then the value of l_{qest} must not be less than L. In strategy 1, the maximum detectable queue length is L, and the absolute error is $\Delta l/2$. When the intersection entrance has more than one lane and each lane is deployed with EMS according to layout strategy 1 (Figure 16(b)), the l_{qest} of each lane can be obtained by Eq. (7). The queue length of the intersection entrance is the average l_{qest} of all lanes, and the corresponding absolute error is $\Delta l/2$ as well.

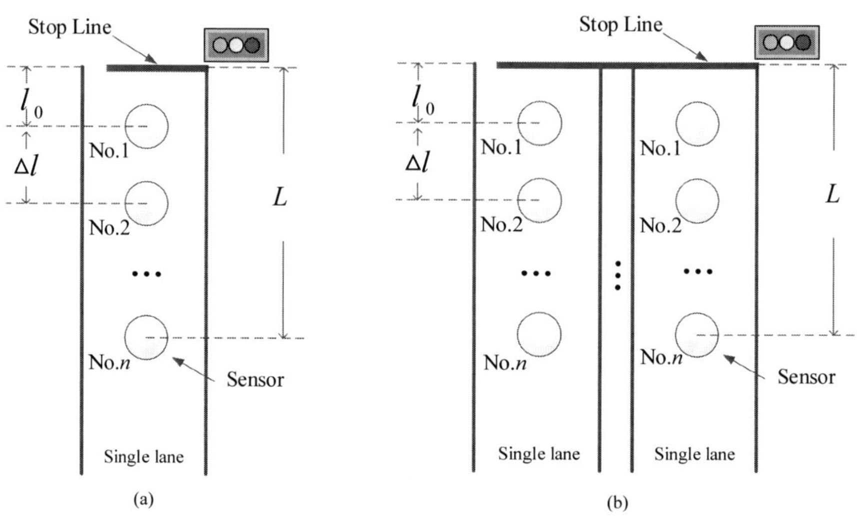

Figure 16. Layout of strategy 1 (a, single lane; b, multiple lanes).

4.3.2. Layout Strategy 2: Two-Point Deployment

This layout strategy deploys two sensors within a designated distance in a lane, which is regarded as layout strategy 2 (Figure 17). The distance between the 1st sensor and the stop line is l_0, and the distance between the 1st and 2nd sensor is l_{12}. For layout strategy 2, l_{qest} can be obtained by Vehicle Matching-based Method (VMM) or Deductive Analysis-based Method (DAM). VMM obtains l_{qest} by matching the signal structures of different vehicles. During a measuring time interval, the vehicles indexed $i = 1, 2, \ldots, N$ cross the 2nd sensor at times $t_1 < t_2 < \ldots < t_N$. The vehicles indexed $j = 1, 2, \ldots, M$ cross the 1st sensor at times $s_1 < s_2 < \ldots < s_M$. The 2nd sensor measures the 'signature' X_i of each passing vehicle i and records the corresponding time t_i. The 1st sensor measures the 'signature' Y_j of each passing vehicle j and records the corresponding time s_j. Thus the outputs of these two sensors consist of two arrays $\mathbf{D}_1=\{(s_j, Y_j), j= 1, \ldots, M\}$ and $\mathbf{D}_2=\{(t_i, X_i), i= 1, \ldots, N\}$. If the 2nd sensor vehicle i is the same as the 1st sensor vehicle j, we will denote it as $X_i \rightarrow Y_j$ according to vehicle matching algorithm. The queuing time of vehicle j detected by the 1st sensor can be shown as follows:

$$T_q(j)=s_j-t_i-\Delta \tag{8}$$

where $T_q(j)$ is the queuing time of vehicle j; and Δ is the average time interval of free driving of a vehicle from the 2nd sensor to the 1st sensor.

Under continuous traffic flow, if there is no lane-changing or lagging behavior, the vehicles will pass the 2nd sensor and the 1st sensor in this order. Then the queuing time of vehicle k passing the 1st sensor is $T_q(k)=(T_k-t_k-\Delta)$. If $T_q(k)>^{\text{TM}}$, the vehicle k is a valid queued vehicle. Typically, the value ^{TM} should be 0. Practically, the value ^{TM} can be greater than 0 to increase the robustness of the method. The number of queued vehicles n_q can be obtained by comparing the vehicles passing through the 1st and the 2nd sensor in order during a signal cycle. Meanwhile, the l_{qest} of the signal cycle can be calculated by Eq. (9):

$$l_{qest} = \sum_{i=1}^{n} l_i + (n_q - 1) l_{inter} \quad (9)$$

where l_i is the length of the ith vehicle; and l_{inter} is the average queue distance between queued vehicles.

The 1st or 2nd sensor can be treated as a cross-section sensor when each direction of the intersection entrance has more than one lane. In this case, $\mathbf{D}_1$ and $\mathbf{D}_2$ each record vehicle information of several lanes in the same direction, and VMM obtains the number of queued vehicles. When queuing, drivers always automatically reach the queue length equilibrium among lanes in the same direction. Then, based on Eq. (9), l_{qest} can still be obtained by computing the average number of queued vehicles in f each lane.

DAM is a deductive inference method used to obtain l_{qest}. In Figure 17(a), we denote the start time as t_0 when no vehicle exists between the 1st and the 2nd sensor. When the vehicle flow passing from the 2nd to the 1st sensor is increasing, if the 1st sensor detects a stopped vehicle at t_1, it means that the traffic light turns to red light and the vehicles begin to queue. And if the 1st sensor detects the first passing vehicle at t_2, it means the traffic light turns to green and the queued vehicles begin to dissipate. Denote the number of vehicles passing the 2nd sensor as $n_{2\#}$ and the 1st sensor as $n_{1\#}$ during t_0 to t_2, if the 2nd sensor doesn't detect a stopped vehicle at t_2, then the number of queued vehicles n_q during this signal cycle is $n_q = n_{2\#} - n_{1\#}$, and the corresponding l_{qest} is shown in Eq. (10).

$$l_{qest} = \sum_{i=1}^{n_{2\#} - n_{1\#}} l_i + (n_{2\#} - n_{1\#} - 1) l_{inter} \quad (10)$$

Otherwise, if the 2nd sensor detects a stopped vehicle at t_2, l_{qest} must not be less than $l_0 + l_{12}$, as shown in Eq. (11).

$$l_{qest} \geq l_0 + l_{12} \quad (11)$$

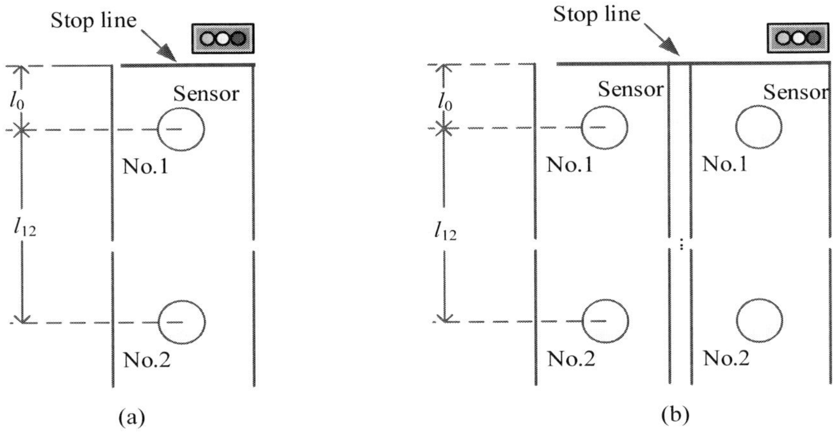

Figure 17. Layout strategy 2 of two-point deployment (a, single lane; b, multiple lanes).

In Figure 17(b), the 2nd and 1st sensor can also be treated as a cross-section sensor. Assume the number of vehicles passing the 2nd and the 1st sensor are $n'_{2\#}$ and $n'_{1\#}$ during t_0 to t_2, respectively, if the 2nd sensor doesn't detect a stopped vehicle at t_2, the corresponding l_{qest} is expressed in Eq. (12); otherwise, l_{qest} is shown in Eq. (11).

$$l_{qest} = \frac{\sum_{i=1}^{n'_{2\#}-n'_{1\#}} l_i + \left(n'_{2\#} - n'_{1\#} - 1\right) l_{inter}}{n_{lane}} \tag{12}$$

where n_{lane} is the number of lanes in one direction at the intersection entrance.

When an intersection entrance is the two-direction type, there will be different cases according to the type of signal control mechanism. If all of the lanes are controlled, the estimation method is the same as the one-direction type. In this section, only two cases, two-direction controlled separately (Figure 18(a)) and one-direction controlled (typically left-turn lane is controlled but right-turn lane not, Figure 18(b)), will be discussed. For sharing a same signal phase, the straight-left lane or straight-right lane should be treated as a straight lane and won't be discussed independently in this section.

If the 2nd sensor is deployed within the solid lane line, it can be assumed that no lane changing behavior exists between the 1st and the 2nd sensor. In that case, VMM and DAM can be used to obtain l_{qest}. Otherwise, if lane changing behavior exists between the 1st and the 2nd sensor, the vehicle flows in these lanes don't obey the queue length equilibrium. In this case, VMM will be more suitable to estimate l_{qest}. If a vehicle pair $X_i \rightarrow Y_j$ is detected by the sensors in the same lane, the l_{qest} can be estimated by Eqs. (8) and (9); otherwise, an additional time factor of lane changing t_a is calibrated, Eqs. (8) will be rewritten as Eq. (13).

$$T_q(j)=s_j-t_i-\Delta-t_a \tag{13}$$

The two-direction type with one-direction controlled is shown in Figure 18(b). The estimation algorithm is the same as in Figure 18(a) for the controlled direction of lanes. If the number of lanes in each direction are greater than one, the method based on cross-section sensors may also be used. The same condition should be applied for the three-direction type, which will not be discussed specifically in the following texts. Next, we will discuss the conditions with one lane in each direction for the three-direction type.

As mentioned in Section 1, the controlled mechanisms of three-direction type mainly consists of four cases, whose queue lengths can also be estimated using the VMM or DAM. For example, Case 1 can be converted into two-direction type with one-direction controlled; Case 3 can be converted into one-direction type; Case 4 can be converted into two-direction controlled separately. Case 2 will be a little more complicated, as shown in Figure 19. If no lane changing behavior exists between the 1st and the 2nd sensor, the l_{qest} can be obtained by using DAM directly; otherwise, VMM will be more suitable to estimate l_{qest} with an additional time factor t_a yet to be determined.

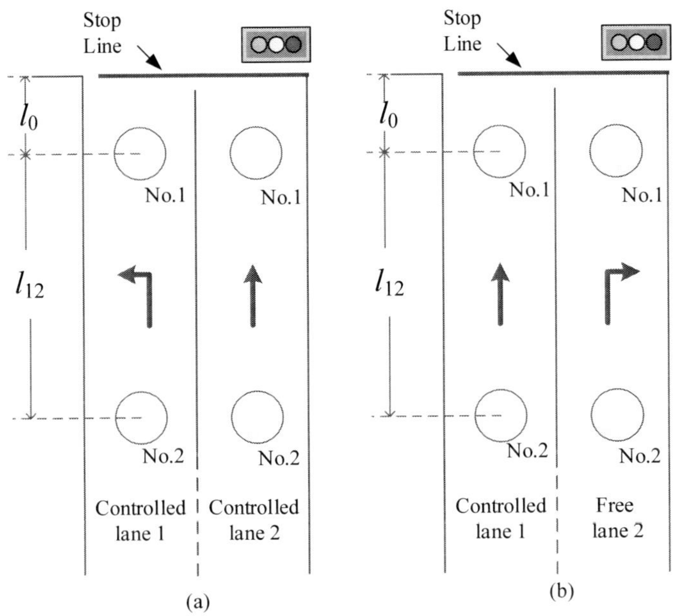

Figure 18. Layout strategy 2 for two-direction types (a, controlled separately; b, one-direction controlled).

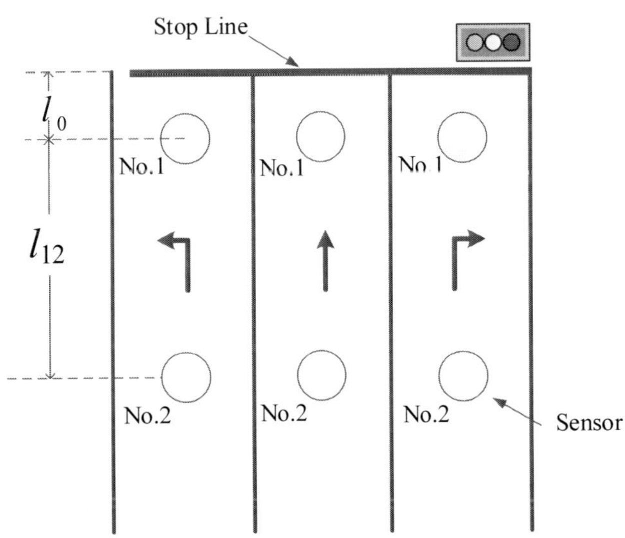

Figure 19. Layout strategy 2 for three-direction type with controlled left-turn and straight lane and an uncontrolled right-turn lane.

4.3.3. Layout Strategy 3: Three-Point Deployment

For layout strategy 3, the 1st sensor is deployed with a distance of l_0 away from the stop line. The 2nd sensor is deployed at the conjunction point of the solid lane line and dotted lane line with a distance of l_{12} away from the 1st sensor. The 3rd sensor is deployed at the dotted lane line with a distance of l_{23} away from the 2nd sensor. The location of the 3rd sensor depends on the maximum value of queue length estimation or distance of to adjacent upstream intersection. Layout strategy 3 combines the computational simplicity of DAM and the improvement of detection accuracy. Its basic theory of queue length estimation is the same with layout strategy 2. The maximum valid value of the queue length estimation is $l_0+l_{12}+l_{23}$. Figure 20 shows the layout strategy 3 for two-direction and three-direction type. According to the location of each sensor, there will be no lane changing behavior between the 1st and 2nd sensor, while there are possible lane changing behaviors between the 2nd and 3rd sensor. The estimation rules will be discussed separately as follows:

1) If the 2nd sensor of a certain lane doesn't detect a stopped vehicle, it shows that the queue length of that lane is less than l_0+l_{12}, and the corresponding estimation algorithm of l_{qest} is shown as Eq. (10).
2) If the 2nd sensor of a certain lane detects a stopped vehicle, it shows that the queue length of that lane is between l_0+l_{12} and $l_0+l_{12}+l_{23}$. The estimation method of the number of stopped vehicles between the 2nd and the 3rd sensor is based on the queue length equilibrium.
3) If the 3rd sensor of a certain lane detects a stopped vehicle, the queue length will be no less than $l_0+l_{12}+l_{23}$.

4.3.4. Layout Strategy 4: Single Sensor Deployment

Layout strategy 4 is the simplest and most cost-optimal strategy among all the strategies above. The single sensor is deployed close to the stop line at a distance of l_0 (Figure 21). The layout strategy 4 cannot obtain real-time queue length, but can estimate queue length of the immediate past signal cycle. Accordingly, strategy 4 can achieve traffic signal optimization and control, basic traffic information acquisition, and LOS evaluation of signalized intersections.

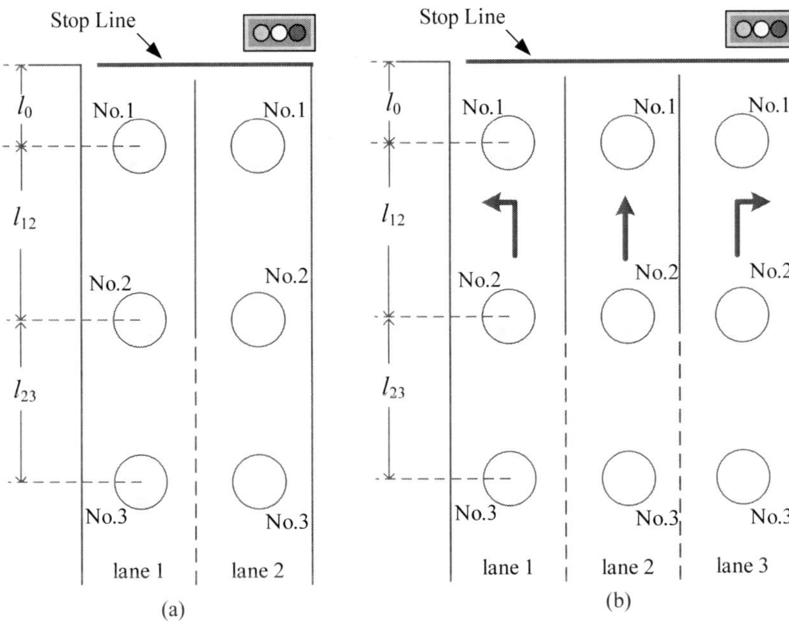

Figure 20. Layout strategy 3 (a, two-direction type; b, three-direction type).

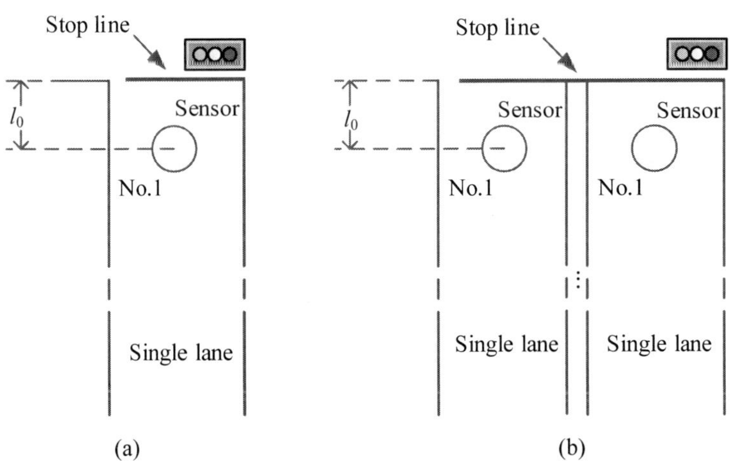

Figure 21. Layout strategy 4 (a, one lane; b, multi-lane).

Layout strategies 1 through 3 can achieve real-time queue length estimation based on DAM. However, because of the use of more sensors, these strategies possess higher costs, have more complicated sensor networks, consume more energy, and require higher precision time synchronization algorithms. Consequently, the single sensor-based queue length estimation method should be studied more to overcome these obstacles.

The process of vehicle queuing complies with the car-following models. Based on car-following behavior, a typical vehicle queuing scenario at a signalized intersection is presented in Figure 22. To simplify the issue, this section makes the following assumptions:

1) The queued vehicles will not change lanes or overtake other vehicles any more. When they are passing through the intersection, they will stay in the same lane and departure the queue in turn.
2) Each queued vehicle satisfies the 'uniform-acceleration-uniform' pattern. A queued vehicle maintains uniform speed (v_{start}) at the start. As v_{start} is low, the speed from 0 to v_{start}, and its corresponding acceleration time can be neglected. If the distance between the vehicle and the front vehicle is d_{need}, the vehicle begins to accelerate. When traveling at an upper limit speed v_{max}, the vehicle begins to keep the uniform speed (v_{max}) again. If the queued vehicle satisfies the acceleration condition when starting to run, it will accelerate to the upper limit speed v_{max} directly.

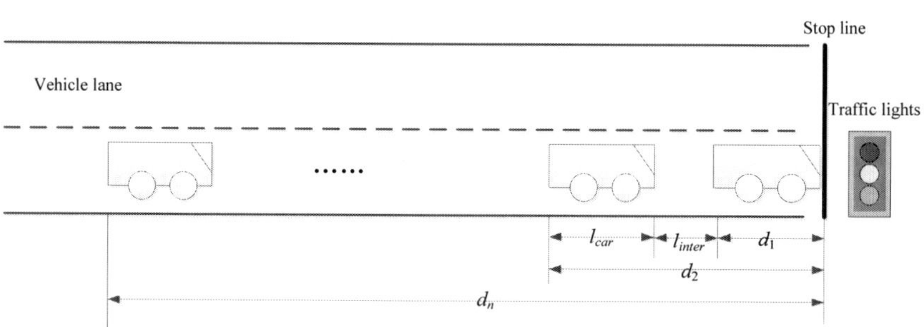

Figure 22. Typical road signal intersection queuing scene.

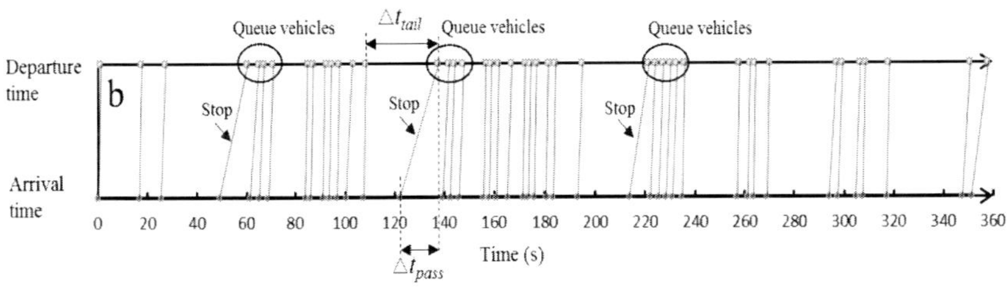

Figure 23. Time pairs of vehicles passing through an EMS in turn.

The arrival and departure time of each vehicle can be obtained using vehicle detection algorithms (Li et al., 2014a). Then, the time pairs of consecutive vehicles passing through an EMS in turn can be got (Figure 23). The time-difference between adjacent front- and back-vehicles and the time-difference between the arrival and departure time of each vehicle are denoted as Δt_{tail} and Δt_{pass}, respectively.

In order to clearly describe the process of queue formation and dissipation, this section also makes the following notations and parameters:

a) The 1st vehicle in the queue lines up to the stop line; the sensor can detect the first vehicle; the average vehicle length is denoted as l_{car}, and accordingly $d_1=l_{car}$.
b) The following distance between two adjacent vehicles is denoted as l_{inter}.
c) The sum of reaction and action time is denoted as t_r.
d) The vehicle acceleration is a and the number of queued vehicles is denoted as N.
e) The uniform speed time of the ith queued vehicle is denoted as $t_c(i)$.
f) The time of the vehicle tail passing the stop line is denoted as $\Delta t_{tail}(i)$ and the speed is denoted as $v(i)$.
g) The time of the vehicle passing the sensor is denoted as $\Delta t_{pass}(i)$.
h) The time difference between the tail of the ith and the $(i\text{-}1)^{\text{th}}$ vehicle passing the stop line is denoted as $\Delta t_{tail}(i)$.
i) The distance between the tail of the ith queued vehicle and the stop line is denoted as d_i, then $d_i=i\times l_{car}+(i-1)l_{inter}$.

Denote the time when the first vehicle in the queue starts to move as 0, then $\Delta t_{pass}(1) = \sqrt{2l_{car}/a}$ and $\Delta t_{tail}(1) = t_{tail}(1)$-0 = $\sqrt{2l_{car}/a}$.

4.3.4.1. Models

4.3.4.1.1. Δt_{tail} Model

The Δt_{tail} model uses the variation rules of $\Delta t_{tail}(i)$ to obtain the number of queued vehicles. $t_{tail}(i)$ should first be determined. $t_{tail}(i)$ is related to the performance of vehicle i-1. The following two cases will be discussed.

Case I: meet the condition of acceleration

In case I, when starting to move, the 2nd vehicle of the queued vehicles satisfies the acceleration condition. The distance between the 2nd and the 1st vehicle is greater than d_{need} ($at_r^2/2 + l_{inter} \geq d_{need}$). Then the 2nd vehicle and the following vehicles will begin to accelerate. By analyzing queue dissipation, it indicates that when N is small, the speed of any vehicle can never reach v_{max} when the vehicle tail reaches the stop line. However, when N is big enough, there must be a vehicle m ($m<N$) in the queue, whose speed will reach v_{max} before its tail leaves the stop line. Then it will begin to keep the uniform speed (v_{max}). The value of m can be obtained through computation or simulation based on the conditions of case I.

If N is small, the speed of vehicle N doesn't reach v_{max} when its tail reaches the stop line, which means $v(N)<v_{max}$, then $\Delta t_{tail}(i)$ can be computed according to Eq. (14):

$$\Delta t_{tail}(i) = \sqrt{2d_i/a} - \sqrt{2d_{i-1}/a} + t_r, \quad i = 2,3,\ldots,N \tag{14}$$

If N is big enough, then $\Delta t_{tail}(i)$ can be obtained according to Eq. (15):

$$\Delta t_{tail}(i) = \begin{cases} \sqrt{2d_i/a} - \sqrt{2d_{i-1}/a} + t_r, & 2 \leq i < m \\ (l_{inter} + l_{car})/v_{max} + t_r, & m \leq i \leq N \end{cases} \tag{15}$$

Case II: did not meet the condition of direct acceleration

In case II, when the 2nd vehicle starts, the distance between the 2nd and 1st vehicle is less than d_{need}, that is $at_r^2/2 + l_{inter} < d_{need}$. The 2nd vehicle and the following vehicles will begin with uniform speed and then accelerate. The time that the 2nd vehicle begins to accelerate needs to be determined.

Similarly, if N is small, the 2nd vehicle and the following vehicles will not satisfy the acceleration condition, which means the queued vehicles keep uniform speed to pass the stop. Accordingly, $\Delta t_{tail}(i)$ can be obtained by Eq. (16). The difference of acceleration time between vehicle i and vehicle i+1 satisfies the relationship in Eq. (17).

$$\Delta t_{tail}(i) = (l_{inter} + l_{car}) / v_{start} + t_r, \qquad i = 2,3,\ldots,N \tag{16}$$

$$t_c(i+1) - t_c(i) = \sqrt{2(d_{need} - l_{inter} - v_{start}t_r)/a} \tag{17}$$

If N is big enough, there will be a vehicle k surely begin to accelerate before its tail reaches the stop line and the speed of vehicle k will not reach v_{max} when its tail passes the stop line. As the number of queued vehicles continues to increase, there must be a vehicle m whose speed will surely reach v_{max} before its tail leaves the stop line. The values of k and m can be obtained through computation or simulation based methods. Then $\Delta t_{tail}(i)$ can be obtained according to Eq. (18).

$$\Delta t_{tail}(i) = \begin{cases} (l_{inter} + l_{car}) / v_{start} + t_r, 2 \le i < k \\ t_c(i) - t_c(i-1) + (v_{tail}(i) + v_{tail}(i-1))/2, k \le i < m \\ (l_{inter} + l_{car}) / v_{max} + t_c(i+1) - t_c(i), m \le i \le N \end{cases} \tag{18}$$

where $v_{tail}(i) = \sqrt{v_{start}^2 + 2a(d_i - v_{start}(t_c(i) - (i-1)t_r))}$, and $v_{tail}(i)$ represents the speed of vehicle i when its tail leaves the stop line.

4.3.4.1.2. Δt_{pass} Model

The Δt_{pass} model applies the variation rules of Δt_{pass} to estimate queue length. Similar to the Δt_{tail} model, two cases will be discussed correspondingly.

Case I: meet the condition of direct acceleration

The condition is similar to the case I of the Δt_{tail} model. If N is small, the speed of each queued vehicle doesn't reach v_{max} when its tail passes the stop line ($v(i)<v_{max}$ (i=2, 3, …, N)). Then $\Delta t_{pass}(i)$ can be obtained as shown in Eq. (19).

$$\Delta t_{pass}(i) = \left(\sqrt{2d_i / a} - \sqrt{2(d_i - l_{car})/a}\right) \tag{19}$$

If N is big enough, the speed of a vehicle m will surely reach v_{max} before its tail leaves the stop line ($v(m)=\sqrt{2a(d_m - l_{car})} \geq v_{\max}$). $\Delta t_{pass}(i)$ can be obtained as shown in Eq. 20).

$$\Delta t_{pass}(i) = \begin{cases} \left(\sqrt{2d_i / a} - \sqrt{2(d_i - l_{car})/a}\right), & 2 \leq i < m \\ l_{car} / v_{\max}, & m \leq i \leq N \end{cases} \tag{20}$$

Case II: did not meet the condition of direct acceleration

Similar to case II of the Δt_{tail} model, if N is small, the queue vehicles will not satisfy the acceleration condition. Then $\Delta t_{pass}(i)$ can be obtained by Eq. (21).

$$\Delta t_{pass}(i)=l_{car}/v_{start} \tag{21}$$

If N is big enough, $\Delta t_{pass}(i)$ can be obtained by Eq. (22).

$$\Delta t_{pass}(i)=\begin{cases} l_{car}/v_{start}, & 2\le i<k \\ l_{car}/\left(\left(v(i)+v_h(i)\right)/2\right), & k\le i<m \\ l_{car}/v_{\max}, & m\le i\le N \end{cases} \tag{22}$$

where $v_h(i)=\sqrt{v_{start}^2+2a\left(d_{i-1}+l_{inter}-v_{start}\left(t_c(i)-(i-1)t_r\right)\right)}$, and $v_h(i)$ means the speed of vehicle i when its head reaches the stop line. The calibration of k and m is similar to the Δt_{tail} model.

During the queue dissipation, the variation of tendency of $\Delta t_{pass}(i)$ is similar to $\Delta t_{tail}(i)$ in each case. It indicates that $\Delta t_{tail}(i)$ and $\Delta t_{pass}(i)$ are not incremental and limited, which are the theoretical basis of the queue length estimation for the immediate past signal cycle. The vehicle waveforms obtained by FMS reflect the changes in the magnetic field. Actually, the arrival and departure time gained by vehicle detection algorithms always differ from the theoretical ones. The arrival time is earlier and the departure time is later than their theoretical time, which will result in an additional length l_e when calibrating the actual lengths of vehicles (Li et al., 2010). Denote $\Delta t_{cpas}(i)$ as the revised value of $\Delta t_{pass}(i)$, then $\Delta t_{cpas}(i)$ is shown in Eq. (23).

$$\Delta t_{cpas}(i)=r\times\Delta t_{pass}(i) \tag{23}$$

where l_e is extra length, r is correction factor and $r=(l_{car}+l_e)/l_{car}$.

The first queued vehicle for each signal cycle in a consecutive traffic flow should first be determined. Two criterions to to determine the first queued vehicle for the immediate past signal cycle is followed:

a) $\Delta t_{tail}(q)+\Delta t_{cpas}(q\text{-}1)\ge t_{red}$. Where q is the vehicle index in a consecutive traffic flow (q=1,2,...), t_{red} is the time duration of red for the queue lane/lanes. The "=" indicates that when the traffic signal turns red, vehicle q-1 just passes through the intersection and vehicle q and its following vehicles start to queue. When the traffic signal turns green, vehicle q starts at the same time followed by each vehicle behind.

b) $\Delta t_{cpas}(q) \geq \Delta t_{cpas}(1)$. Where $\Delta t_{cpas}(1) = r \times \Delta t_{pass}(1)$, which represents the minimum time duration for a queued vehicle. The condition "$\Delta t_{cpas}(q) = \Delta t_{cpas}(1)$" indicates that vehicle q happens to stop for queuing and the signal turns green at that moment.

If vehicle q in a consecutive traffic flow meets the two criterions above, it indicates that vehicle q is the 1st queued vehicle of the immediate past signal cycle. Secondly, the last queued vehicle of the same signal cycle will be identified by the methods proposed in the section below. Finally, the number of queued vehicles of each signal cycle can be obtained.

4.3.5.2. Methods

4.3.5.2.1. Tail Interval-Based Method (TIM)

TIM uses the Δt_{tail} model to estimate queue length. Typically, when the first queued vehicle is identified, the $\Delta t_{tail}(i)$ of the following vehicles can be obtained in turn immediately during the next green and yellow signal. Comparing the two values of $\Delta t_{tail}(i)$ (one is obtained from the EMS; the other is calculated by Eqs. (14-18)), if the relative error of these two values is within a set value δ_1, then it is indicated that the ith vehicle is in the queue; else it is not. The algorithm based on TIM (Algorithm 6) is presented as following:

Algorithm 6: the vehicle queue length estimation algorithm based on TIM	
1:	Determining the first queued vehicle based on the two criterions for the immediate past signal cycle, and let k=1, i=1
2:	**if** no following vehicle exists during the immediate past green and yellow signal **then** go to step 4 **else** $i=i+1$; go to step 3 **end if**
3:	**if** the relative error of these two values of $\Delta t_{tail}(i)$ is within a set value δ_1 **then** $k=k+1$; return to step 2 **else** go to step 4 **end if**
4:	$l_{est}=d_k$; for the next signal cycle, return to step 1

For the sake of simplicity, in step 3, Algorithm 6 can just use some values of $\Delta t_{tail}(i)$ rather than all of them to compute relative errors of the queues vehicles. For example, the front j vehicles can use $\Delta t_{tail}(2)$ to determine whether they are queued in order, and from vehicle j+1 to N, the algorithm can use $\Delta t_{tail}(j+1)$.

4.3.5.2.2. Passing Time-Based Method (PTM)

PTM just uses the Δt_{pass} model. The algorithm of PTM is similar to Algorithm 6 with a slight difference in step 3. The new criterion of PTM in step 3 will be 'the relative error of these two values of $\Delta t_{cpas}(i)$ is without a set value δ_2'. PTM can also just use some values of $\Delta t_{cpas}(i)$. For example, the front j vehicles use $\Delta t_{cpas}(j)$ to determine whether they are queued in order, and the vehicles from j+1 to N can use $\Delta t_{cpas}(j+1)$.

4.3.5.2.3. Tail Interval and Passing Time-Based Method (T-PM)

T-PM takes both TIM and PTM into account and it satisfies both of the criterions of TIM and PTM. Then the criterion of PTM in step 3 will be 'the relative error of these two values of $\Delta t_{tail}(i)$ is within a set value δ_1 **&** the relative error of these two values of $\Delta t_{cpas}(i)$ is without a set value δ_2'. Denote the numbers of queued vehicles obtained by TIM, PTM and T-PM as n_t, n_p, and n_{tp}, respectively, then we have $n_{tp}=\min\{n_t, n_p\}$.

5. EMS Allocation Strategies for Urban Road Networks

5.1. Traffic Spatial Properties with Fixed Magnetic Sensors

In this section, the nonuniformity of the spatial distribution of traffic information based on fixed traffic sensors is presented (Bao et al., 2016a; H. Li et al., 2017). The acquisition of the all-weather transportation information mainly depends on the fixed sensors arranged on the road. These sensors mainly provide a dot or zonal distribution, such as commonly used inductive loop sensors, magnetic sensors, microwave sensors, and video sensors. The scope of the sensors is primarily one type of road (one-lane road, multilane road, or one-way road section). In the actual traffic information collection system, a "one-way road section" will always be the common unit, that is, traffic information from one testing location represents the traffic information from a one-way road section of each lane.

Combined with the space structure of the road network and the space distribution of detection points, this section shows the topology model of an X-road network with fixed traffic sensors as an example in Figure 24. Lines with pointed arrows represent one-way roads, the numbered circles represent road intersections or ramp junctions and the dots on the lines represent the sensors and the layout scheme. The topology model describes the physical structure and the layout position of each sensor in the road network, and serves as the model to describe the spatial distribution of transportation information.

Figure 24 shows that traffic information using the fixed sensors is acquired by using limited information at the points of sensors (POS) in a section of a road or road network to extrapolate traffic information of the entire road or road network. Traffic information at the points with no sensors (PNS) can be estimated using limited information of known POS. Then, every point (POS or PNS) in the road networks will be a value or extent of credibility of traffic information.

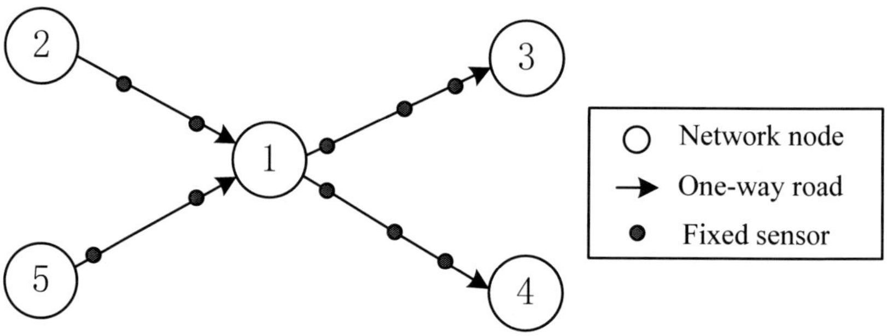

Figure 24. Topology model of an X-road network with fixed traffic sensors.

Here, the concept of traffic information credibility (TIC) is defined, which can be regarded as an information index with limited POS in road networks. As shown in our previous work (Li et al., 2015) and (Li et al., 2014b), the value of TIC is high at POS and may be lower at PNS. TIC represents the spatial distribution of information credibility derived from fixed traffic sensors. As early as 1997, Drezner and Wesolowsky (Drezner and Wesolowsky, 1997) indicated that the probability of an event detected by the fixed sensors is a decreasing function of the distance from the sensor points. The authors postulated a general probability function of distance: $\pi(d)$, which is monotone decreasing with the increase of distance. Traffic information acquired by existing methods have the spatial imbalance in the road section or road network. TIC is proposed to establish an information measure to describe this spatial imbalance. As shown in (Li et al., 2015), there are several necessary demands to use the new information index of TIC in traffic networks. Sometimes, a decision maker only cares about the traffic information at the PNS, where the traffic information cannot be obtained directly and is estimated by the information from the POS. Currently, it is very important to know the contribution or credibility of the POS to the PNS. Furthermore, TIC reveals the spatial distribution characteristic of traffic information from fixed sensors, which may show a new way to analyze the dynamic distribution of traffic information from mobile sensors (moving GPS, mobile phone, floating car, etc.).

By adding the values of TIC in traffic networks, the two-dimensional network topology model can be expanded to a three-dimensional spatial distribution TIC model. In Figure 25, the two-dimensional network space consists of the intersection, the one-way road, and the traffic sensors. TIC consists of the third dimension, which is perpendicular to the two-dimension network space, and the value at one point of the two-dimension network represents the network information credibility of that point. It is assumed that: (1) There is an association or causality between the upstream traffic state and downstream traffic state within certain distance. (2) Traffic information of the PNS in the section or road network can be estimated by the POS, and the estimate has credibility to a certain extent. (3) The closer the PNS and the POS are to each other, the more credible the traffic information estimate of the PNS. The value of TIC at POS is set equal to 100%, regardless of the accuracy of the sensors. The value of TIC at the PNS can be estimated using the information from the PNS, and may have a value of 100% or less. As the distance between the PNS and POS increases, the information credibility of the PNS decreases.

Li et al. (Li et al., 2015, 2014b) proposed the sensor information credibility function (ICF) to describe the three-dimensional spatial distribution of TIC derived from a sensor along a one-way road, which is similar to the π(*d*) function proposed by in (Drezner and Wesolowsky, 1997). The sensor ICF sets the location of the sensor as the origin of coordinates and is described as follows:

$$\mathrm{ICF}(x)=f(x)\ x\in(-\infty,+\infty) \tag{24}$$

where x is the coordinate on the one-way road with traffic sensors.

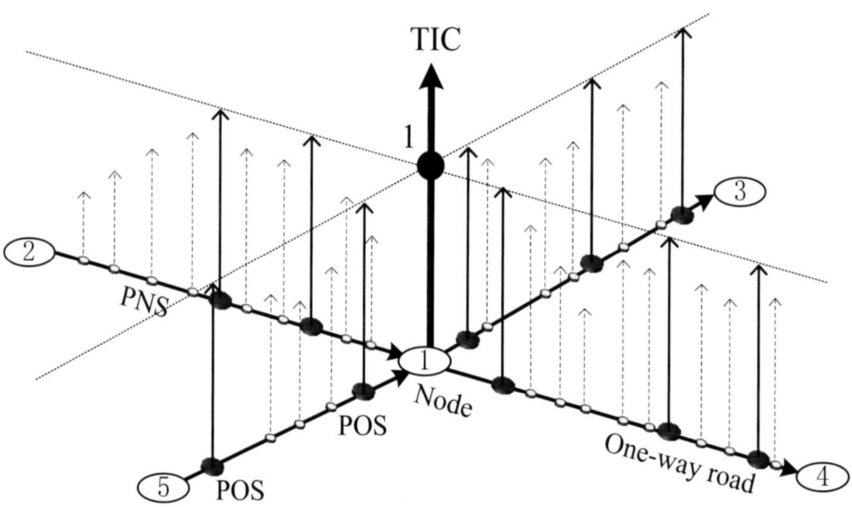

Figure 25. Three-dimension spatial distribution model of TIC from part of the X-road network.

A positive value indicates that the point is downstream of traffic flow relative to the sensor, and a negative value indicates that the point is upstream. ICF(x) is the value of TIC at the coordinate x. f(x) is the function expression of ICF and is dependent on different road conditions and sensor locations. There are some characteristics of $f(x)$: (1) for any given x, $0 \leq f(x) \leq 1$; (2) $f(0)=1$, $f(\infty)=0$; and (3) the degree of the absolute contribution(d_{ac}) and the degree of the relative contribution(d_{rc}) of TIC of the sensor between x_1 and x_2 are defined as follows:

$$d_{ac}(f, x_1, x_2) = \int_{x_1}^{x_2} f(x)dx \tag{25}$$

$$d_{rc}(f, x_1, x_2) = \frac{\int_{x_1}^{x_2} f(x)dx}{\int_{-\infty}^{+\infty} f(x)dx} \tag{26}$$

ICF describes the spatial distribution of TIC both upstream and downstream of the sensor on a one-way road. Furthermore, to accommodate a road or road network with multiple sensors, Li et al. (Li et al., 2014b) proposed the space superposition methods of two ICF functions. The information credibility function of a one-way road (RICF) and the information credibility function of a traffic network (NICF) are used to describe the TIC spatial distribution of a one-way road or a road network with fixed sensors located in it, respectively. The definitions are as follows:

$$\text{RICF} = \bigoplus_{i=1}^{n} \text{ICF}_i \tag{27}$$

$$\text{NICF} = \bigotimes_{k=1}^{m} \text{ICF}_k \tag{28}$$

where i is sensor #i of the one-way road; k is sensor #k of the road network; n is the total number of sensors on the one-way road; m is the total number of sensors on the road network; $\oplus$ is the ICF spatial overlay operator of the one-way road; and $\otimes$ is the ICF spatial overlay operator of the road network.

Li et al. (Li et al., 2014b) also presented two common spatial overlay modes of ICFs: the maximum mode and the information fusion mode. The maximum mode selects the maximum ICF of each overlay as the road ICF of that point. This mode is based on our tendency to accept the higher creditable or reliable information and to reject the lower creditable or reliable one. This mode is easy to operate and has practical significance. The information fusion mode acquires more abundant traffic information, provides more accurate original traffic information, and in general results in higher creditable information. This mode is more complex to operate, but generates more comprehensive information.

Based on the maximum mode and information fusion mode, Eq. (27) can be presented as:

$$\mathrm{RICF}(x) = \bigoplus_{i=1}^{n} \mathrm{ICF}_i(x) = \max{}_{i=1}^{n}(\mathrm{ICF}_i(x_i)) \tag{29}$$

$$\mathrm{RICF}(x) = \bigoplus_{i=1}^{n} \mathrm{ICF}_i(x) = 1 - \prod{}_{i=1}^{n}(1 - \mathrm{ICF}_i(x_i)) \tag{30}$$

where x_i is the distance between point x and sensor i. The other parameters are the same as in Eq. (27).

5.2. Sensor Allocation Optimization for Urban Road Networks

5.2.1. Methodology

ICF can represent different spatial distributions of traffic information credibility. Different micro-regions, such as ramp entrances, ramp exits, merging areas, or diversion areas, will have different spatial characteristics of traffic flow. One can conclude that the spatial distributions of traffic information credibility will be multiform. In fact, for different conditions of roads and sensors, there are various $f(x)$ that can calibrate the ICF. As discussed in (Li et al., 2014b), three typical ICFs (exponential attenuation ICF (EAF), linear attenuation ICF (LAF), and step attenuation ICF (SAF)) are shown in Figure 26. Based on field data, Li et al. (Li et al., 2015, 2014b) have determined that the ICF can be calibrated using field data. Studying these three typical ICFs will be helpful to reveal more complex spatial distributions of traffic information credibility for actual road networks.

As shown in Figure 26(a), the EAF represents a nonuniform gradient of traffic information credibility that can be formulated by an exponential function, which stands for an exponentially decreasing probability of detection. As discussed in (Li et al., 2015, 2014b), k, k', a, a' and p_1, p_2, p'_1, p'_2, q_1, q'_1 are the coefficients of these three typical ICFs and the values of all these coefficients are positive.

Most sensor allocation models and methods in traffic information acquisition are proposed for specific applications. As noted by Hu et al.(S. R. Hu et al., 2009), most existing methods focus on the subset problems, such as traffic OD estimation, travel time estimation, and road connectivity, which will not be suitable to analyze the influence pattern of the sensor allocation and spatial distribution of traffic information derived from fixed sensors. Three new measures (ICF, RICF, and NICF) are proposed to effectively obtain comprehensive information at POS and PNS. Based on the optimization model proposed by Li et al. (Li et al., 2015), the minimum investment model (MIM) of sensor allocation is designed from the perspective of probability and statistics to achieve minimum investment cost. In the MIM model, from the starting point to the terminal point of the research road, n sensors of candidate layout points are deployed uniformly and densely. If the length of the road is L, the distance between the adjacent candidate points is d, where $d=L/(n-1)$. The value of d determines the accuracy of the MIM solution, in which the accuracy error is not greater than $d/2$. A larger number of candidate layout points can be set to improve the model accuracy. In engineering applications, this model uses the off-line optimization. When results can be obtained within an acceptable time period, it is not necessary to consider the algorithm complexity, and the model accuracy can be improved by setting a small value of d. The parameter d can be determined based on the error criterion in practical engineering, and the original number of the candidate layout points is obtained by rounding up equation $n=L/d+1$.

If the starting point of the one-way road is the origin of coordinate o, the expression formula of the MIM model is as follows:

$$\min \quad z=\sum_{i=1}^{n}(X_i \times C(i)) \tag{31}$$

$$s.t.\begin{cases} X_i \in \{0,1\} \\ x_i=(i-1)d \\ d_{rc}(f_i(x-x_i),x_i,a_{i+}) \le d_{rc}(f_i(x),0,\max(f_i^{-1}(\frac{D_{\min}}{Q(i)}))) \\ d_{rc}(f_i(x-x_i),a_{i-},x_i) \le d_{rc}(f_i(x),\min(f_i^{-1}(\frac{D_{\min}}{Q(i)})),0) \end{cases} \tag{32}$$

where z is the total investment. x_i is the coordinate of candidate point i, and x_{i+1}-x_i=d=$L/(n$-1). f_i is the sensor information reliability function of candidate point i. The f_i^{-1} is the inverse function of f_i; X_i is 0 or 1, where X_i=0 indicates that the candidate point i is the point of no sensor, and x_i=1 indicates the point of the sensor. $C(i)$ is the total sensor cost of candidate point i, including the sensor equipment cost and costs of installation and maintenance. $Q(i)$ is the sensor accuracy of candidate point i. $D_{\min}$ is the minimal information credibility allowed in engineering, and the value is to ensure the completeness and effectiveness of the one-way road traffic information acquisition. a_{i-}, and a_{i+} are for the function RICF abscissa of the left and right minimum points at candidate point i, while X_i=1, a_{1-}=0, and a_{n+}=L.

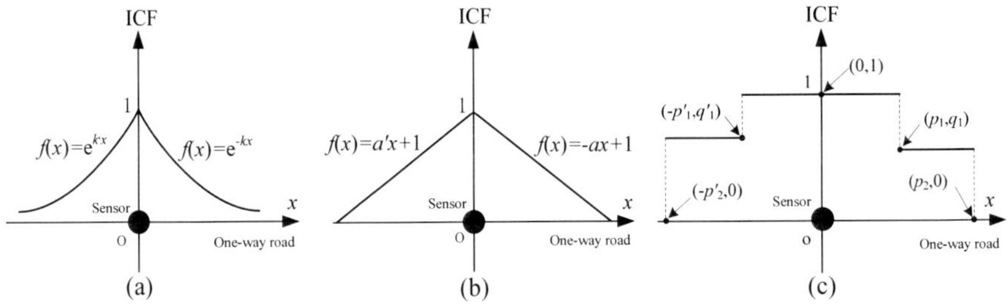

Figure 26. Three typical ICFs (a, The EAF; b, The LAF; c, The two-SAF).

Similar models have been utilized in the literature (Li et al., 2014b) in which the shortest path is searched using graph theory. While useful, this method has limitations in that sensors must be deployed at the starting point and end point of the research section, thus it cannot determine the optimal distance between the starting point of the research section and the first sensor, and the optimal distance between the end point of the research section and the last sensor, under the optimal sensor layout condition. A universal solution of the MIM model is proposed by introducing the generalized shortest path matrix.

The MIM can be considered as an n variable 0-1 integer programming model. According to the model structure, the time complexity is $O(2^n)$ if using the enumeration method to solve the model. Using the directed graph G constructed in the preliminary study of the paper, the time complexity can be reduced from $O(2^n)$ to $O(n^3)$.

Each vertex of the directed graph G represents the initial sensor candidate layout point, and the numbers 1, 2…, n are for the serial numbers of the initial sensor candidate layout points of the one-way road, that is, they are the serial numbers of each candidate layout sensor. The directed line e_{ij} ($i, j \in \{1, 2, ..., n\}$ and $i<j$) which begins at the low numbered vertex and ends at the high number vertex is the edge of the directed graph G.

Definition 1: the equivalency credibility D_{ij} of the directed line e_{ij}($i, j \in \{1, 2, ..., n\}$ and $i<j$):

$$D_{ij} = \begin{cases} D_{\min} & \min(f_i(a_{ij} - x_i)Q(i), f_j(a_{ij} - x_j)Q(j)) \geq D_{\min} \\ j-i & \min(f_i(a_{ij} - x_i)Q(i), f_j(a_{ij} - x_j)Q(j)) < D_{\min} \end{cases} \quad (33)$$

where a_{ij} is the spatial overlay coordinate of the two ICFs of sensor i and sensor j, that is,

$$f_i(a_{ij}\text{-}x_i)=f_j(a_{ij}\text{-}x_j) \quad (34)$$

Definition 2: the weight w_{ij} of the directed line e_{ij} ($i, j\in\{1, 2, ..., n\}$ and $i<j$):

$$w_{ij} = \begin{cases} D_{ij} & i < j \\ n & i \geq j \end{cases} \quad (35)$$

There are three steps to obtain z_m:

1) Denote $\mathbf{W}_{n\times n}$ as the weight matrix of the directed graph G ($\mathbf{W}(i,j)=w_{ij}$ ($i, j\in\{1, 2, ..., n\}$)), using the Floyd algorithm in graph theory, which addresses situations with negative weights, to calculate the shortest path matrix $\mathbf{P}_{n\times n}$ of two arbitrary points in the directed graph G.
2) The generalized shortest path matrix $\mathbf{P}'_{n\times n}$ can be calculated by the following equations:

$$\mathbf{P}'(i, j) = \begin{cases} \mathbf{P}(i, j) + D_i + D_j & i < j \\ \mathbf{P}(i, j) & i \geq j \end{cases} \quad (36)$$

where

$$D_i = \begin{cases} 0 & f_i(-x_i)Q(i) \geq D_{\min} \\ i-1 & f_i(-x_i)Q(i) < D_{\min} \end{cases},$$

$$D_j = \begin{cases} 0 & f_j(L - x_j)Q(j) \ge D_{\min} \\ n - j & f_j(L - x_j)Q(j) < D_{\min} \end{cases}.$$

3) Make $\mathbf{P}'(i_{min}, j_{min})=\min(\mathbf{P}'_{n\times n})$, and denote $\mathbf{P}_V$ as the vertex set of $\mathbf{P}'(i_{min}, j_{min})$, then, the elements of $\mathbf{P}_V$ are the optimal sensor locations of the MIM. The distance between adjacent vertices in $\mathbf{P}_V$ is the optimal sensor spacing. Then, the minimum comprehensive investment z_m is calculated as below:

$$z_m = \sum_{i\in \mathbf{P_V}} C(i) \tag{37}$$

In practical engineering, the model parameters can be similar or identical when the study objects share the same property, the research network is small, the road grade is the same, and/or the sensors are the same types. The differences in the model parameter among the different locations is very small for small regions or the same section. In this case, a simplified MIM model (SMIM) can be given as follows:

$$z(D_{\min}) = nC = \left(\frac{L - d_1 - d_2}{\max(f^{-1}(\frac{D_{\min}}{Q})) + (-\min(f^{-1}(\frac{D_{\min}}{Q})))} + 1 \right) C \tag{38}$$

where f is the ICF of each POS; C is the total cost of each sensor; Q is the accuracy of each sensor; d_1, d_2 are the distance parameters of the road endpoints, that is, d_1 is the distance between the first sensor required in engineering and the starting point of the one-way road or the upstream intersection, and d_2 is the distance between the last sensor required in engineering and the ending point of the one-way road or the downstream intersection. Other parameters are the same as in Eq. (31) and Eq. (32).

Eq. (38) stipulates that the objective function is the function of $D_{\min}$ when the roads and types of sensor are determined. If d_l and d_2 are not constrained in engineering, when Eq. (38) obtains the optimum solution, the distance between two adjacent sensors is $d=d_l+d_2$, and d_l and d_2 must meet the following conditions:

$$d_1 = -\min(f^{-1}(\frac{D_{\min}}{Q}) \quad , \quad d_2 = \max(f^{-1}(\frac{D_{\min}}{Q}) \tag{39}$$

Specifically, when the left and right of $f(x)$ are symmetrical, d_l and d_2 need to meet the following conditions:

$$d_1 = d_2 = \frac{d}{2} = \max(f^{-1}(\frac{D_{\min}}{Q}) = -\min(f^{-1}(\frac{D_{\min}}{Q}) \tag{40}$$

Now, Eq. (38) can be transformed into the following equation:

$$z(D_{\min}) = nC = \left(\frac{L - d_1 - d_2}{2\max(f^{-1}(\frac{D_{\min}}{Q}))} + 1 \right) C = \frac{LC}{2\max(f^{-1}(\frac{D_{\min}}{Q}))} \tag{41}$$

5.2.2. Urban Traffic Network Application

This section uses the SMIM to verify the effectiveness and practicability of the aforementioned method for a regional road network in Beijing consisting of a 2nd ring expressway and six major inner arterial roads (two vertical and four horizontal). The regional road network shown in Figure 27(a) can be converted into the one-way section-based topology shown in Figure 27(b). Each of the one-way sections is divided by the network nodes (signal intersections or ramp junctions). Each serial number of each section is shown in Figure 27(b), and the basic information of each section is shown in Table 9.

Table 9. Basic information of each one-way road of the regional road network (H. Li et al., 2017)

Index	Road name	Length (km)	Index	Road name	Length (km)	Index	Road name	Length (km)
O1	Outer 2nd Ring	2.7	I1	Inner 2nd Ring	2.7	S1	Xinjiekou Street	1.8
O2	Outer 2nd Ring	3.8	I2	Inner 2nd Ring	3.8	S2	Xisi Street	2.8
O3	Outer 2nd Ring	3.1	I3	Inner 2nd Ring	3.1	S3	Xuannei Street	0.8
O4	Outer 2nd Ring	2.8	I4	Inner 2nd Ring	2.8	S4	Xuanwai Street	1.2
O5	Outer 2nd Ring	0.9	I5	Inner 2nd Ring	0.9	S5	Caishikou Street	2.1
O6	Outer 2nd Ring	1.2	I6	Inner 2nd Ring	1.2	S6	Yonghegong Street	1.8
O7	Outer 2nd Ring	4.5	I7	Inner 2nd Ring	4.5	S7	Dongsi Street	2.8
O8	Outer 2nd Ring	4	I8	Inner 2nd Ring	4	S8	Chongnei Street	0.8
O9	Outer 2nd Ring	4.4	I9	Inner 2nd Ring	4.4	S9	Chongwai Street	0.9
O10	Outer 2nd Ring	1.6	I10	Inner 2nd Ring	1.6	S10	Tiantan Road	2.5
O11	Outer 2nd Ring	0.9	I11	Inner 2nd Ring	0.9	N1	Xinjiekou Street	1.8
O12	Outer 2nd Ring	2.8	I12	Inner 2nd Ring	2.8	N2	Xisi Street	2.8
E1	Pinganlixi Street	1.5	W1	Pinganlixi Street	1.5	N3	Xuannei Street	0.8
E2	Zhang-Di Street	3.8	W2	Di-Zhang Street	3.8	N4	Xuanwai Street	1.2
E3	Dongsishitiao	1.5	W3	Dongsishitiao	1.5	N5	Caishikou Street	2.1
E4	Funei Street	1.5	W4	Funei Street	1.5	N6	Yonghegong Street	1.8
E5	Changan Street	3.8	W5	Changan Street	3.8	N7	Dongsi Street	2.8
E6	Jiannei Street	1.5	W6	Jiannei Street	1.5	N8	Chongnei Street	0.8
E7	Xuanwumenxi Street	1.8	W7	Xuanwumenxi Street	1.8	N9	Chongwai Street	0.9
E8	Chong-Xuan Street	3.8	W8	Xuan-Chong Street	3.8	N10	Tiantan Road	2.5
E9	Chongwen-mendong Street	1.5	W9	Chongwen-mendong Street	1.5			
E10	Guangnei Street	2.2	W10	Guangnei Street	2.2			
E11	Zhu-Luo Street	3.8	W11	Luo-Zhu Street	3.8			
E12	Guangqumennei Street	2.2	W12	Guangqumennei Street	2.2			

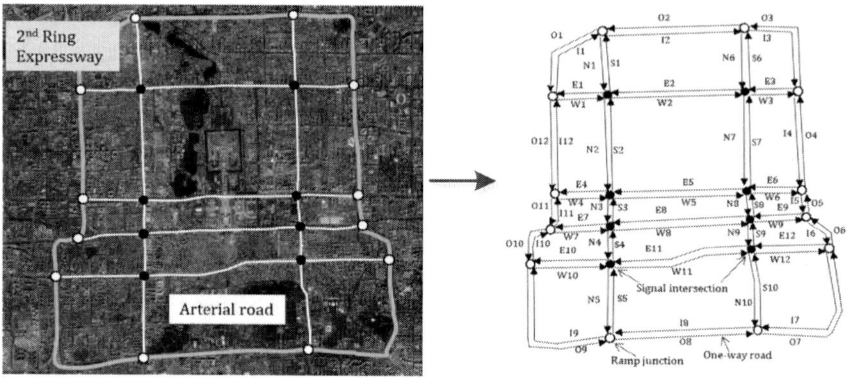

(a) (b)

Figure 27. Road network (a, the road network of the example; b, one-way road based topology model).

From Li et al. (Li et al., 2014b), the ICF in the urban expressway presents a negative exponent attenuation as the distance increases in Beijing, which can be expressed as:

$$f(x)=e^{-k|x|} \tag{42}$$

where k is the attenuation coefficient, and $k \geq 0$.

Denote k_1 as the attenuation coefficient of the 2nd ring expressway, and k_2 as the attenuation coefficient of the six major roads. Traffic flows on the typical urban arterial roads with signalized intersections such as those six roads in this example are often not continuous, and the fluctuation of traffic flow on these roads will be greater than on the 2nd ring expressway, therefore, k_2 will always be greater than k_1.

In addition, as Li et al. (Li et al., 2014b) revealed that for different parameters (such as volume and speed) of traffic flow, their corresponding spatial distributions of TIC are different. That is, their attenuation coefficients will be uneven. Let k_1^{vol} and k_1^{spe} denote the attenuation coefficients of volume and speed of traffic flow on the 2nd ring expressway, as observed in Li et al. (Li et al., 2014b), k_1^{vol} =0.1592, k_1^{spe} =0.2043. Denote k_2^{vol} and k_2^{spe} as the attenuation coefficients of volume and speed on the arterial roads respectively, it is reasonable to set k_2^{vol} =0.20 and k_2^{spe} =0.25.

Because all the intersections are the key points of the regional road network (Figure 27), it is of priority to place sensors in each intersection. Sensors at 12 ramp connection points and 8 signal intersections are deployed first. The actual number of sensors depends on the number of cross sections, the size of the intersection, and the detecting region of each sensor (single lane, multi-lane, one section, or multi-section). In this paper, the number of sensors to deploy at each key point (network node) is assumed as 1, then, a total of 20 sensors are needed for all key points. The sensor layout strategy of the sections will be determined between these key points. It is assumed that the minimum information credibility D_{min} that practical engineering allows is 0.90, the detection accuracy Q of each sensor for any traffic flow parameter is 95%, and D_{min} will always be less than Q.

Here, substitute k_1^{vol}, k_1^{spe}, k_2^{vol}, k_2^{spe} into Eq. (42) and the sensor ICFs for different traffic network conditions are obtained. Then by substituting the ICFs and D_{min}=0.90 and Q=0.95 into Eq. (40), the recommended sensor spacings for different network conditions can be calculated. Denote d_e^{vol}, d_e^{spe}, d_a^{vol}, d_a^{spe} as the optimal sensor spacings for traffic volume information acquisition on the 2nd ring expressway, traffic speed information acquisition on the 2nd ring expressway, traffic volume information acquisition on the six arterial roads and traffic speed information acquisition on the six arterial roads, respectively, then d_e^{vol}=0.679 km, d_e^{spe}=0.529 km, d_a^{vol}=0.541 km, and d_a^{spe}=0.433 km. These are the recommended sensor spacings for the four conditions in the example. By using the length information of the road network in Table 9, the number of the sensors required on each section (except the key points of the network) are calculated. Table 10 shows the optimal sensor number of each one-way road in the road network for traffic volume information acquisition, and Table 11 shows the similar information for traffic speed information acquisition. In Table 10 and Table 11, the calculated values are obtained by dividing the length of each section by the optimal distance between adjacent sensors. The number of sensors must be an integer, and the number of sensors at network nodes has been calculated individually, so the sensor number values are obtained by rounding the calculated values up to the nearest integer and subtracting 1.

Table 10. Optimal sensor number of each one-way road in the road network for traffic volume information acquisition (H. Li et al., 2017)

Index	Calculated value	Sensor number	Index	Calculated value	Sensor number	Index	Calculated value	Sensor number
O1	3.975	3	I1	3.975	3	S1	3.329	3
O2	5.595	5	I2	5.595	5	S2	5.179	5
O3	4.564	4	I3	4.564	4	S3	1.480	1
O4	4.122	4	I4	4.122	4	S4	2.219	2
O5	1.325	1	I5	1.325	1	S5	3.884	3
O6	1.767	1	I6	1.767	1	S6	3.329	3
O7	6.625	6	I7	6.625	6	S7	5.179	5
O8	5.889	5	I8	5.889	5	S8	1.480	1
O9	6.478	6	I9	6.478	6	S9	1.665	1
O10	2.356	2	I10	2.356	2	S10	4.624	4
O11	1.325	1	I11	1.325	1	N1	3.329	3
O12	4.122	4	I12	4.122	4	N2	5.179	5
E1	2.774	2	W1	2.774	2	N3	1.480	1
E2	7.028	7	W2	7.028	7	N4	2.219	2
E3	2.774	2	W3	2.774	2	N5	3.884	3
E4	2.774	2	W4	2.774	2	N6	3.329	3
E5	7.028	7	W5	7.028	7	N7	5.179	5
E6	2.774	2	W6	2.774	2	N8	1.480	1
E7	3.329	3	W7	3.329	3	N9	1.665	1
E8	7.028	7	W8	7.028	7	N10	4.624	4
E9	2.774	2	W9	2.774	2			
E10	4.069	4	W10	4.069	4			
E11	7.028	7	W11	7.028	7			
E12	4.069	4	W12	4.069	4	Total sensor number		238

From Table 10 and Table 11, the total optimal number of sensors in the road network for different information acquisition requirements is obtained. By adding these 20 sensors at network nodes (all intersections and ramp connection points), for traffic volume information acquisition the total number of the recommended sensors is 258, and for traffic speed information acquisition the total number of the recommended sensors is 326. If the sensor type for each link or node in the road network has been determined in practical projects, the model parameter $C(i)$ or C will be obtained. Finally, the minimum investment can be determined according to Eq. (37) or Eq. (41).

To analyze the sensitivity of model parameters, Figure 28 demonstrates the influence of the attenuation coefficient k on the optimal sensor spacing d (H. Li et al., 2017). For a determined D_{min}, the value of TIC attenuates along with the distance slower for smaller values of k, which results in increased distance between the sensors, and fewer sensors need to be deployed, which ultimately lowers investment cost. If k is less than 0.1, the optimal sensor spacing d is more sensitive to changes in the value of k. In practice, efforts to reduce the k value in engineering can effectively reduce the number of sensors needed, which lowers the total investment cost. Common recommended practices include the design of an isolation area to ease the interference of motorized vehicles and non-motorized vehicles, improvement of the pedestrian facilities to reduce the conflict between pedestrians and motorized vehicles, separation of passenger and cargo traffics to reduce traffic flow velocity difference, or development of connected and automated vehicles to establish cruise vehicle fleets.

Figure 29 demonstrates the influence of the sensor accuracy Q on the optimal sensor spacing d (H. Li et al., 2017). It is shown that a higher accuracy Q of the sensor and a greater value of d results in a fewer investment cost. To improve the traffic information detection accuracy, new traffic sensor technology should be developed that is more cost-effective, results in less loss of data and has lower communications errors caused by construction or telecommunication. Finally, from Figure 28 and Figure 29, if the required D_{min} value is lower, d will be larger, resulting in less sensors being deployed and lower investment costs. In engineering, a lower D_{min} value can be achieved to take further advantage of the raw data in transportation, develop data mining technology, and improve the utilization rate of existing data.

Table 11. Optimal sensor number of each one-way road in the road network for traffic speed information acquisition (H. Li et al., 2017)

Index	Calculated value	Sensor number	Index	Calculated value	Sensor number	Index	Calculated value	Sensor number
O1	5.101	5	I1	5.101	5	S1	4.161	4
O2	7.179	7	I2	7.179	7	S2	6.473	6
O3	5.857	5	I3	5.857	5	S3	1.850	1
O4	5.290	5	I4	5.290	5	S4	2.774	2
O5	1.700	1	I5	1.700	1	S5	4.855	4
O6	2.267	2	I6	2.267	2	S6	4.161	4
O7	8.502	8	I7	8.502	8	S7	6.473	6
O8	7.557	7	I8	7.557	7	S8	1.850	1
O9	8.313	8	I9	8.313	8	S9	2.081	2
O10	3.023	3	I10	3.023	3	S10	5.780	5
O11	1.700	1	I11	1.700	1	N1	4.161	4
O12	5.290	5	I12	5.290	5	N2	6.473	6
E1	3.468	3	W1	3.468	3	N3	1.850	1
E2	8.785	8	W2	8.785	8	N4	2.774	2
E3	3.468	3	W3	3.468	3	N5	4.855	4
E4	3.468	3	W4	3.468	3	N6	4.161	4
E5	8.785	8	W5	8.785	8	N7	6.473	6
E6	3.468	3	W6	3.468	3	N8	1.850	1
E7	4.161	4	W7	4.161	4	N9	2.081	2
E8	8.785	8	W8	8.785	8	N10	5.780	5
E9	3.468	3	W9	3.468	3			
E10	5.086	5	W10	5.086	5			
E11	8.785	8	W11	8.785	8			
E12	5.086	5	W12	5.086	5	Total sensor number		306

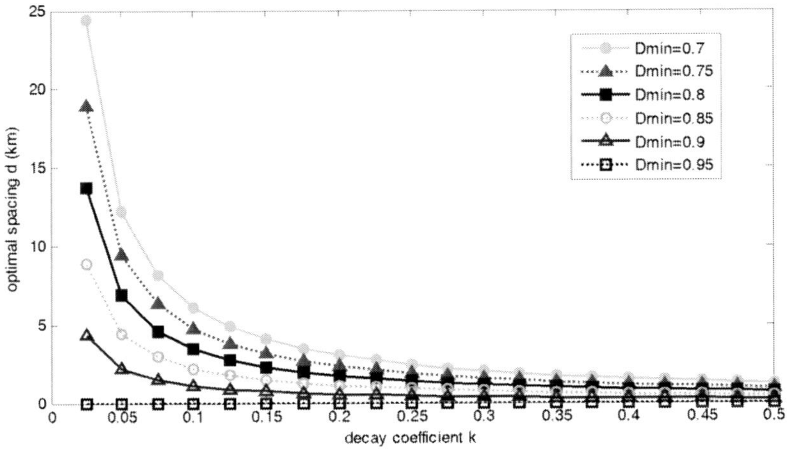

Figure 28. Sensitivity curves of *d-k* with different D_{min}.

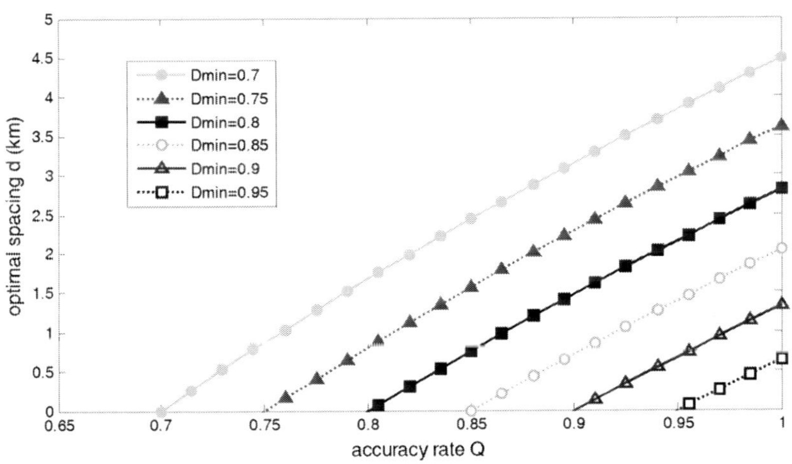

Figure 29. Sensitivity curves of *d-Q* with different D_{min}.

CONCLUSION

A low-cost and energy-efficient type of multi-function wireless traffic EMS was designed and developed. There are several advantages of EMS that are ideal for large-scale deployment and time-sustainable detection for traffic information acquisition. With the development of sensor and network technologies, a type of hybrid sensor network has been formed in traffic information acquisition systems.

Based on multi-function wireless traffic EMS and corresponding hybrid sensor network, this chapter proposed a vehicle detection algorithm based on EMS and puts forward improved vehicle detection algorithms using magnetic sensors under traffic jam conditions. The basic vehicle detection algorithm with less computational complexity was introduced for vehicle counting under low traffic volume conditions. And the improved vehicle detection algorithms were found to be efficient and accurate vehicle detection methods to address the drift problem caused by weather environments and the "tailgating effect" caused by traffic flow environments. As a low-cost, energy-efficient and is easily installed, traffic EMS is suitable for large-area deployment and all-weather, all-day, traffic surveillance.

Traffic state acquisition methods with different layout strategies of EMS were introduced. These methods could obtain necessary traffic information such as vehicle speed, vehicle classification and queue length for traffic state estimation. According to different sensor layout strategies, speed-acquisition-oriented Vehicle Length-based and Time Difference-based detection methods are discussed. If the length of each vehicle is precisely measured, the Vehicle Length based method will improve in accuracy. For vehicle classification, a decision tree model based on CART algorithm was used to execute on-line vehicle classification in EMS nodes. The comparison results of current methods also showed that the proposed method has some advantages in aspects of accuracy rate, sample robustness and execution time. For queue length estimation at signalized intersections, this chapter modeled and analyzed the queue length estimation mechanisms by different layout strategies. The layout strategy within a single EMS was found to be the most economical strategy to estimate queue length of the immediate past signal cycle. The experimental results also showed that the proposed single-sensor based method is simple and economical, and will be suitable for large-scale deployment and application.

Finally, sensor location problem based on fixed sensors (such as EMS) for urban road networks was studied. The heterogeneity of spatial distribution of information credibility from sensor networks was analyzed and a new measure, information credibility function, was proposed to describe this heterogeneity and to analyze traffic spatial properties with fixed magnetic sensor locations. The proposed method generates recommended sensor layout strategies for different information acquisition requirements of traffic flow. Experimental results show that this method possess the following attributes: simplicity, practicability and maneuverability. Further studies should be focused on the following aspects: to explore ICF in more detail and to calculate ICF for more road type networks for different information acquisition and to be more detailed in sensor layout problems for traffic control, traffic queue detection at intersections etc., which can improve the level of fine traffic management.

References

Bajwa, R., Rajagopal, R., Varaiya, P., Kavaler, R., Street, N., 2011. In-Pavement Wireless Sensor Network for Vehicle Classification. *Proc. 10th ACM/IEEE Int. Conf. Inf. Process. Sens. Networks* 85–96.

Bao, X., Li, H., Qin, L., Xu, D., Ran, B., Rong, J., 2016a. Sensor Location Problem Optimization for Traffic Network with Different Spatial Distributions of Traffic Information. *Sensors* 16, 1790. doi:10.3390/s16111790.

Bao, X., Li, H., Xu, D., Jia, L., Ran, B., Rong, J., 2016b. Traffic vehicle counting in jam flow conditions using low-cost and energy-efficient wireless magnetic sensors. *Sensors* (Switzerland) 16, 1–16. doi:10.3390/s16111868.

Burresi, G., Giorgi, R., 2015. A Field Experience for a Vehicle Recognition System using Magnetic Sensors, in: *4th Mediterranean Conference on Embedded Computing*. Budva, Montenegro, pp. 178–181.

Cai, Q., Wang, Z., Guo, X., Wu, B., 2013. New Calculating Method for HCM 2000 Queue Length Estimation Procedures with the Application of Floating Car Data. *Procedia-Social Behav. Sci.* 96, 2201–2210.

Cai, Q., Wang, Z., Zheng, L., Wu, B., Wang, Y., 2014. Shock Wave Approach for Estimating Queue Length at Signalized Intersections by Fusing Data from Point and Mobile Sensors. *Transp. Res. Rec. J. Transp. Res. Board* 2422, 79–87. doi:10.3141/2422-09.

Cetin, M., 2012. Estimating Queue Dynamics at Signalized Intersections from Probe Vehicle Data. *Transp. Res. Rec. J. Transp. Res. Board* 2315, 164–172. doi:10.3141/2315-17.

Cheung, S., Coleri, S., Dundar, B., Ganesh, S., Tan, C.-W., Varaiya, P., 2005. Traffic Measurement and Vehicle Classification with Single Magnetic Sensor. *Transp. Res. Rec.* 1917, 173–181. doi:10.3141/1917-19.

Cheung, S., Varaiya, P., 2007. Traffic Surveillance by Wireless Sensor Networks: *Final Report,* California Path Program Institute of Transportation Studies University of California, Berkeley. Berkeley.

Coifman, B., Kim, S.B., 2009. Speed estimation and length based vehicle classification from freeway single-loop detectors. *Transp. Res. Part C Emerg. Technol.* 17, 349–364. doi:10.1016/j.trc.2009.01.004.

Comert, G., 2013. Simple analytical models for estimating the queue lengths from probe vehicles at traffic signals. *Transp. Res. Part B Methodol.* 55, 59–74. doi:10.1016/j.trb.2013.05.001.

Daubaras, A., Zilys, M., 2012. Vehicle detection based on magneto-resistive magnetic field sensor. *Elektron. ir Elektrotechnika* 2, 27–32. doi:10.5755/j01.eee.118.2.1169.

Drezner, Z., Wesolowsky, G.O., 1997. On the best location of signal detectors. *IIE Trans.* 29, 1007–1015. doi:10.1080/074081797 08966419.

Gontarz, S., Szulim, P., Senko, J., Dybala, J., 2015. Use of magnetic monitoring of vehicles for proactive strategy development. *Transp. Res. Part C Emerg. Technol.* 52, 102–115. doi:10.1016/j.trc. 2014.12.007.

Goratti, L., Haapola, J., Ieee, M., Kato, S., Ieee, F., 2012. Highly Reliable Star and Sub-Mesh Hybrid Sensor Network for Smart Grid Monitoring. *IEEE Globecom Work.* 1–6.

Han, B., Zhang, K., Yu, X., Kwon, E., Ou, J., 2011. Nickel particle-based self-sensing pavement for vehicle detection. *Meas. J. Int. Meas. Confed.* 44, 1645–1650. doi:10.1016/j.measurement.2011.06.014.

Haoui, A., Kavaler, R., Varaiya, P., 2008. Wireless magnetic sensors for traffic surveillance. *Transp. Res. Part C Emerg. Technol.* 16, 294–306. doi:10.1016/j.trc.2007.10.004.

Harlow, C., Peng, S., 2001. Automatic vehicle classification system with range sensors. *Transp. Res. Part C Emerg. Technol.* 9, 231–247. doi:10.1016/S0968-090X(00)00034-6.

He, Y., Du, Y., Sun, L., Wang, Y., 2015. Improved waveform-feature-based vehicle classification using a single-point magnetic sensor. *J. Adv. Transp.* 49, 663–682. doi:10.1002/atr.1299.

He, Z., Zhu, H., Yu, F., 2014. A vehicle detection algorithm based on wireless magnetic sensor networks, in: *2014 4th IEEE International Conference on Information Science and Technology. IEEE*, pp. 727–730. doi:10.1109/ICIST.2014.6920580.

Honeywell International Inc., 2008. 1- and 2-Axis Magnetic Sensors HMC1001/1002/1021/1022 manual. [WWW Document]. *Honeywell Int. Inc.* URL http://www51.honeywell.com/aero/common/documents/myaerospacecatalog-documents/Missiles-Munitions/HMC_1001-1002-1021-1022_Data_Sheet.pdf (accessed 8.19.08).

Hu, S. R., Peeta, S., Chu, C. H., 2009. Identification of vehicle sensor locations for link-based network traffic applications. *Transp. Res. Part B Methodol.* 43, 873–894. doi:10.1016/j.trb.2009.02.008.

Hu, W., Bulusu, N., Chou, C. T., Jha, S., Taylor, A., Tran, V. N., 2009. Design and evaluation of a hybrid sensor network for cane toad monitoring. *ACM Trans. Sens. Networks* 5, 1–28. doi:10.1145/1464420.1464424.

Iwasaki, Y., Misumi, M., Nakamiya, T., 2013. Robust vehicle detection under various environmental conditions using an infrared thermal camera and its application to road traffic flow monitoring. *Sensors* 13, 7756–7773. doi:10.3390/s130607756.

Jo, Y., Jung, I., 2014. Analysis of vehicle detection with WSN-based ultrasonic sensors. *Sensors* 14, 14050–14069. doi:10.3390/s140814050.

Kaewkamnerd, S., Pongthornseri, R., Chinrungrueng, J., Silawan, T., 2009. Automatic vehicle classification using wireless magnetic sensor. *2009 IEEE Int. Work. Intell. Data Acquis. Adv. Comput. Syst. Technol. Appl.* 420–424. doi:10.1109/IDAACS.2009.5342949.

Kafai, M., Bhanu, B., 2012. Dynamic bayesian networks for vehicle classification in video. *IEEE Trans. Ind. Informatics* 8, 100–109. doi:10.1109/TII.2011.2173203.

Kwong, K., Kavaler, R., Rajagopal, R., Varaiya, P., 2010. Real-time measurement of link vehicle count and travel time in a road network. *IEEE Trans. Intell. Transp. Syst.* 11, 814–825. doi:10.1109/TITS.2010.2050881.

Kwong, K., Kavaler, R., Rajagopal, R., Varaiya, P. P., 2009. Arterial Travel Time Estimation Based on Vehicle Re-Identification Using Wireless Magnetic Sensors. *Transp. Res. Part C* 17, 586–606.

Lamas-Seco, J., Castro, P., Dapena, A., Vazquez-Araujo, F., 2015. Vehicle Classification Using the Discrete Fourier Transform with Traffic Inductive Sensors. *Sensors* 15, 27201–27214. doi:10.3390/s151027201.

Lan, J., Xiang, Y., Wang, L., Shi, Y., 2011. Vehicle detection and classification by measuring and processing magnetic signal. *Meas. J. Int. Meas. Confed.* 44, 174–180. doi:10.1016/j.measurement.2010.09.044.

Li, H., 2014. *Multi-parameter Sensing and Sensor Network Optimization for Road Traffic Information Acquisition*. Beijing Jiaotong University.

Li, H., Chen, N., Qin, L., Jia, L., Rong, J., 2017. Queue length estimation at signalized intersections based on magnetic sensors by different layout strategies. *Transp. Res. Procedia* 25, 1629–1647. doi:10.1016/j.trpro.2017.05.212.

Li, H., Dong, H., Jia, L., Ren, M., 2014a. Vehicle classification with single multi-functional magnetic sensor and optimal MNS-based CART. *Meas. J. Int. Meas. Confed.* 55, 142–152. doi:10.1016/j.measurement.2014.04.028.

Li, H., Dong, H., Jia, L., Ren, M., Li, S., 2015. Analysis of Factors that Influence the Sensor Location Problem for Freeway Corridors. *J. Adv. Transp.* 49, 10–28.

Li, H., Dong, H., Jia, L., Ren, M., Qin, Y., 2014b. A new measure for evaluating spatially related properties of traffic information credibility. *J. Cent. South Univ.* 21, 2511–2519.

Li, H., Dong, H., Jia, L., Xu, D., Qin, Y., 2011. Some practical vehicle speed estimation methods by a single traffic magnetic sensor, in: *IEEE Conference on Intelligent Transportation Systems, Proceedings, ITSC.* Washington, USA, pp. 1566–1573. doi:10.1109/ITSC.2011.6083076.

Li, H., Jia, L., Dong, H., Qin, Y., Xu, D., Sun, X., 2010. Study on spacing optimization for traffic flow detector, in: *IEEE Conference on Intelligent Transportation Systems, Proceedings, ITSC. IEEE,* Madeira Island, pp. 593–598. doi:10.1109/ITSC.2010.5625045.

Li, H., Qin, L., Chang, X., Rong, J., Ran, B., Jia, L., 2017. Sensor layout strategy and sensitivity analysis for macroscopic traffic flow parameter acquisition. *IET Intell. Transp. Syst.* 11, 1–10. doi:10.1049/iet-its.2016.0297.

Liepins, M., Severdaks, A., 2013. Vehicle detection using non-invasive magnetic wireless sensor, in: *2013 21st Telecommunications Forum Telfor (TELFOR).* pp. 601–604. doi:10.1109/TELFOR.2013.6716302.

Lin, W. H., Kulkarni, A., Mirchandani, P., 2003. Arterial travel time estimation for advanced traveler information systems. *Transp. Res. Board 82nd Annu. Meet.*

Liu, H. X., Ma, W., Wu, X., Hu, H., 2012. Real-time estimation of arterial travel time under congested conditions. *Transportmetrica* 8, 87–104. doi:10.1080/18128600903502298.

Liu, H. X., Wu, X., Ma, W., Hu, H., 2009. Real-time queue length estimation for congested signalized intersections. *Transp. Res. Part C Emerg. Technol.* 17, 412–427. doi:10.1016/j.trc.2009.02.003.

Marshall, S. V., 1978. Vehicle detection using a magnetic field sensor. *IEEE Trans. Veh. Technol.* 27, 65–68. doi:10.1109/T-VT.1978.23725.

Martin, P. T., Feng, Y., Wang, X., 2003. *Detector Technology Evaluation.* Utah.

Medina, J., Ramezani, H., Benekohal, R., 2013. Evaluation of Microwave Radar Vehicle Detectors at a Signalized Intersection under Adverse Weather Conditions. *Transp. Res. Rec. J. Transp. Res. Board* 2356, 100–108. doi:10.3141/2356-12.

Nazlibilek, S., Ege, Y., Kalender, O., Sensoy, M. G., Karacor, D., Sazli, M.H., 2012. Identification of materials with magnetic characteristics by neural networks. *Meas. J. Int. Meas. Confed.* 45, 734–744. doi:10.1016/j.measurement.2011.12.017.

Negri, P., 2014. Estimating the queue length at street intersections by using a movement feature space approach. *Iet Image Process.* 8, 406–416. doi:10.1049/iet-ipr.2013.0496.

Oh, C., Ritchie, S.G., 2007. Recognizing vehicle classification information from blade sensor signature. *Pattern Recognit. Lett.* 28, 1041–1049. doi:10.1016/j.patrec.2007.01.010.

Oh, S., Ritchie, S., Oh, C., 2002. Real-Time Traffic Measurement from Single Loop Inductive Signatures. *Transp. Res. Rec. J. Transp. Res. Board* 1804, 98–106. doi:10.3141/1804-14.

Paulraj, M. P., Adom, A. H., Sundararaj, S., Rahim, N. B. A., 2013. Moving Vehicle Recognition and Classification based on Time Domain Approach. *Procedia Eng.* 53, 405–410. doi:10.1016/j.proeng.2013.02.053.

Razakarivony, S., Jurie, F., 2016. Vehicle detection in aerial imagery: A small target detection benchmark. *J. Vis. Commun. Image Represent.* 34, 187–203. doi:10.1016/j.jvcir.2015.11.002.

Sanchez, R. O., Horowitz, R., Varaiya, P., 2011. Analysis of Queue Estimation Methods Using Wireless Magnetic Sensors. *Transp. Res. Rec. J. Transp. Res. Board* 2229, 34–45. doi:10.3141/2229-05.

Shenoy, S., 2005. Simultaneous localization and mobile robot navigation in a hybrid sensor network, in: *2005 IEEE/RSJ International*

Conference on Intelligent Robots and Systems. pp. 1636–1641. doi:10.1109/IROS.2005.1545194.

Sifuentes, E., Casas, O., Pallas-Areny, R., 2011. Wireless Magnetic Sensor Node for Vehicle Detection with Optical Wake-up. *IEEE Sens. J.* 11, 1669–1676.

Taghvaeeyan, S., Rajamani, R., 2014. Portable roadside sensors for vehicle counting, classification, and speed measurement. *IEEE Trans. Intell. Transp. Syst.* 15, 73–83. doi:10.1109/TITS.2013.2273876.

Wang, Q., Zheng, J., Xu, B., Huang, Y., 2015. Analysis and Experiments of Vehicle Detection with Magnetic Sensors in Urban Environments, in: *IEEE International Conference on Cyber Technology in Automation, Control, and Intelligent Systems (CYBER).* Shenyang, China, pp. 71–75.

Wang, R., Zhang, L., Xiao, K., Sun, R., Cui, L., 2014. EasiSee: Real-Time Vehicle Classification and Counting via Low-Cost Collaborative Sensing. *IEEE Trans. Intell. Transp. Syst.* 15, 414–424. doi:10.1109/TITS.2013.2281760.

Xia, Y., Shi, X., Song, G., Geng, Q., Liu, Y., 2014. Towards improving quality of video-based vehicle counting method for traffic flow estimation. *Signal Processing* 120, 1–10. doi:10.1016/j.sigpro. 2014.10.035.

Xiang, Y., Piedrahita, R., Dick, R. P., Hannigan, M., Lv, Q., Shang, L., 2013. A Hybrid Sensor System for Indoor Air Quality Monitoring, in: *2013 IEEE International Conference on Distributed Computing in Sensor Systems. IEEE,* Cambridge, MA, pp. 96–104. doi:10.1109/DCOSS.2013.48.

Yang, B., Lei, Y., 2015. Vehicle detection and classification for low-speed congested traffic with anisotropic magnetoresistive sensor. *IEEE Sens. J.* 15, 1132–1138. doi:10.1109/JSEN.2014.2359014.

Zhang, H., Liu, X., Wang, X., Huo, M., 2008. Principles and applications of magneto-resistance effect. *J. Harbin Inst. Technol.* 40, 362–366.

Zhang, W., Tan, G. Z., Shi, H. M., Lin, M. W., 2010. A distributed threshold algorithm for vehicle classification based on binary proximity sensors and intelligent neuron classifier. *J. Inf. Sci. Eng.* 26, 769–783.

Zheng, J., Ma, X., Wu, Y. J., Wang, Y., 2013. Measuring Signalized Intersection Performance in Real-Time with Traffic Sensors. *J. Intell. Transp. Syst.* 17, 304–316. doi:10.1080/15472450.2013.771105.

In: Advances in Engineering Research
Editor: Victoria M. Petrova
ISBN: 978-1-53614-218-1

Chapter 2

VORTEX AND HIGHLY TURBULENT FLOWS IN THE CHEMICAL ENGINEERING: A REVIEW OF THE SPECIALIZED SOFTWARE PRODUCTS CLUSTER TO PERFORM CALCULATIONS FOR THE GRANULATION DEVICES

Artem Artyukhov*[*] *and Nadiia Artyukhova
Processes and Equipment of Chemical and Petroleum-Refineries Department, Sumy State University, Sumy, Ukraine

ABSTRACT

The study presents the main advantages of the vortex and highly turbulent flows and their implementation area in the chemical engineering. The technological scheme, which includes devices with active hydrodynamic regimes used in the granulation processes, is

[*] Corresponding Author Email: a.artyukhov@pohnp.sumdu.edu.ua.

presented in this work. It also outlines the theoretical bases for the technological line of the devices with active hydrodynamic regimes. Based on the original mathematical models, a set of applied programs is investigated to perform calculations for the granulation device. The mathematical model includes the main hydrodynamic equations for the one- and two-phase flows, thermodynamic equations, the equations of the thermal kinetics and dehydration of the granules etc. All the equations in the mathematical model are adapted for the devices with active hydrodynamic regimes. The study provides a comprehensive overview of the previous author's research results in the sphere of the automation of the granulation devices. The presented software products allow users to perform optimization calculations for the main stages of the granulation process. The obtained calculation results enable to select an optimal construction of equipment with minimal costs for the construction and provision of the technological parameters in the process.

In order to calculate the granulation and final drying stages the software products Vortex Flow©, Conical Channel©, Vortex Granulator©, Classification in vortex flow©, Multistage fluidizer© are created. The modeling of an exhaust gas cleaning process is conducted with the Vortex Tray©. All the software products allow transferring of the calculation results and making the Granulation Unit© cluster, which is demonstrated in this review for the first time.

The methodology to calculate the main operational stages of the granulation device is presented in the work. It is based on the calculation results, which were carried out with the use of the author's software products. The experimental and industrial production line of the porous ammonium nitrate, which is projected with the automated calculations by the Granulation Unit© cluster, is described.

Keywords: software products cluster, granulation, optimization

INTRODUCTION

Vortex single phase and multiphase flows are widely used in the heat transfer and mass transfer processes in chemical technologies [1]. The progress of the target process is intensified several times in response to the high relative velocity of the movement of phases [2]. The three-dimensional vortex flows in the units enable an intense mix, increase a phase contact area and contribute to the establishment of the uniform temperature profile in a working space [3]. The mentioned factors provide an opportunity to decrease the time taken by the heat and mass transfer process in the unit. The constructive solutions, committed to ensure vortex flows, are successfully implemented in the processes of classification and separation [4], heat transfer [5], absorption [6], rectification [7], drying [8, 9], etc.

It should be noted that increasing of the turbulization degree (a vortex movement intensity) is not proportional to the intensification of the target process and the productivity of the unit. Some excess vortexing may lead to the hydrodynamic operational instability (e.g., a return vortex formation, an unintended overlapping of vortexes).

During the heat and mass transfer processes in the two-phase (gas-solid) and three-phase (gas-liquid-solid) systems, the excess intensification of the vortexing may lead to the destruction of the solid phase (via scuffling, cracking, splitting, etc.). Thus, the search for the optimal operation mode for the unit with such working systems becomes topical.

Despite the wide use of the vortex flow technologies, they are just partially implemented in some heat and mass transfer processes. The fluidized bed granulation is among these processes. This relates to the complicated hydrodynamic conditions in such units and to the multiphase system as well. The main challenges of the vortex flows implementation in the fluidized bed granulation are as follows [14-17]:

- ensure the formation of the spherical drops avoiding their large distortion in the gas flow;
- ensure a high level of monodisperse granules of the commercial fraction;
- minimal mixing of the different granule fractions, preventing the collision of the granules' trajectories;
- ensure a stable operation of the fluidized bed;
- a wide range of the load ratio share in the gas, liquid and solid phases;
- the uniform filming of the granules surface;
- the minimal destroy of the granule core;
- ensure the necessary time to heat and dry the granules;

- the formation of the directed outlet for the granules in the unit;
- the allocation of the separated operational areas with various stages of the target process within the working space;
- the implementation of the directed internal circulation of the fines;
- exhaust gas purification immediately in the granulator;
- a multistage use of the heating agent.

These challenges must be addressed through the development of the new organizational forms of the vortex flows in the granulator space, including units of the vortex type. The author's works [18-30] present main solutions on how to decrease the destroying impact on the vortex fluidized bed granulation. These are:

- the height-variable cross-section of the granulator working space ensures various hydrodynamic conditions for the granules of different fractions to move (a motion trajectory, the residence time, a geometric location in the work space);
- the internal circulation of the fines in the annulus of the granulator leads to the enlarging of the granules to the sizes of the commercial fraction by own substandard granules;
- the work space zoning (separate zones for the granule heating, filming of the granule's surface, granule drying, classification of the granules by sizes, separation of the fine granules) ensures the necessary residence time for the granules of the certain size in the work space;
- the work space sectioning by vertical perforated shelves gives a staged change of the temperature and moisture of granules, a multiuse of the heat transfer agent (this constructive solution may also be used in the separate unit for the final drying of granules as depicted below);
- the formation of several zones of the vortex motion of granules and zones with different intensity of the vortex gas flow provides strict fractioning of the granules;

- the use of the vortex mass transfer and separation sections (both inbuilt and mobile) – utilization of the heat and waste products like ammonia dust and steam.

Despite all the presented constructive solutions, which ensure a new organization of the flow motion (within the single unit or the whole technological installation), a search for the optimal constructive and technological approach to the granulation process remains a complex problem.

The review of the modern software products (simulation modeling systems) used in the calculations for the chemical production plants (Aspen HYSYS®, PRO/II, GasCondOil, ChemCAD, DESIGN II, SIGMAFINE, gPROMS ModelBuilder, GIBBS and others) [31-38] shows that such software is mainly implemented for the technological calculation of the specified construction and under certain technological parameters of the chemical plants. The optimization calculation of the technological process requires a multiple initial provisioning; a quantity of such provisions for the optimization calculations may be numerous. The simulation software used for the hydrodynamic and heat-mass transfer processes in the chemical production (ANSYS Fluent, COSMOS FlowWorks, FlowVision, Flow-3D CFD, Ansys CFX, COMSOL Multiphysics and others) [39-44] enable to model only one construction variant of the installation and, thus, it reduces the optimization calculation performance. In practice, we use various models of the numerical solution of the classical equation of hydrodynamics and heat-mass transfer for turbulent flows. These models are used in certain cases, and they have their advantages and disadvantages [45]: Direct numerical simulation (DNS), Reynolds Averaged Navier-Stokes (RANS), Large Eddy Simulation (LES).

Herewith, it should be noted that simulation modeling systems and software used for the hydrodynamic and heat-mass transfer processes in the chemical production are effective for the calculation of the units with specified parameters and tools of the certain design. Nevertheless, the world's practice proves that a computer modeling of the hydrodynamics and heat-mass transfer in the chemical production units may successfully replace the manual experiment. This is depicted in the works [46-55] on the mass transfer columns, the heat transfer units etc.

For this reason, it is necessary to develop the automated complex (a software package) of the optimization calculation for the vortex granulators and supplement equipment (stages of purification and drying) by several optimization criteria (e.g., a minimal workspace volume, a minimal consumption of the heating agent, minimal residence time of granules in the unit while ensuring the quality criteria). Herewith, all the constructive solutions and the software package itself are used to design the residence time steering mechanism for the granules in the vortex granulator's work space.

This work will present the software package of the Granulation Unit© cluster. The cluster is used to make calculations for the porous ammonium nitrate (PAN) production units in the vortex granulators.

Background

The PAN production process is very sensitive to the hydrodynamic and thermodynamic parameters [56, 57]. The main requirements applied for the PAN are a forked network of nanopores of different sizes and a firm core of the granule [21]. A long residence time of the granules in a work space causes damages from the intense collision of granules with each other and a wall of the unit. The process becomes even more complicated when a strong thermal tension appears in the granules if they are overheated [58]. The lack of the residence time may also cause damages to the granules from the excess moisture. As for the nanopores formations, under a long heat treatment the porous structure is mainly formed by the "mechanical" pores (that appears in response to the mechanical damage of the granules and the thermal tension) [59]. Apart from the decreasing fraction strength, this type of pores doesn't allow long restraining of the diesel oil (as a necessary component of the industrial explosive). A short heat treatment causes formation of the poor quantity of the so-called "modification" pores (that are formed in response to the change of the crystal lattice structure and the distance between lattice sites, as well as through the water vaporization and a discharge of the vaporized ammonium on the granules' surface) [60, 61].

The above-mentioned requirements underline the importance of the optimization and automation of the calculations for the vortex granulators. The technological scheme of the PAN production in the vortex granulators is depicted in Figure 1. The scheme shows a stage of the heat treatment of the granules after moisturizing by the stepped "mild" dehydration in the inline units with active hydrodynamic modes (a vortex granulator *VG* and a gravitational shelf dryer *GSD* are depicted).

The exhaust gas purification and the heat utilization are provided in a separate unit (an absorber *A* with a vortex mass transfer and separation contact cascade), and in the granulator as well (presented below). The software calculations for the heat and waste utilization are similar in each case.

The description of the tools, for which optimization calculations are subject to the automation, is further presented. The vortex granulator operates (Figure 2) as follows (the work shows only the target process of the granulation and the organization of the flows movement common to all vortex granulators).

The high heating agent is tangentially supplied to the granulator through the relevant sleeve and then it is transferred to the vortex gas-distributing hub. While moving through the hub, the agent rotates about the vertical axis and gets a helical motion. Damp granules are added to the helical motion flows. While contacting with the axisymmetric vortex flows of the heating agent, the granules are subject to the overcut heating with dehydration, formation of the porous surface layer, and drying in the heat flows. The weight of the granules is gradually reducing.

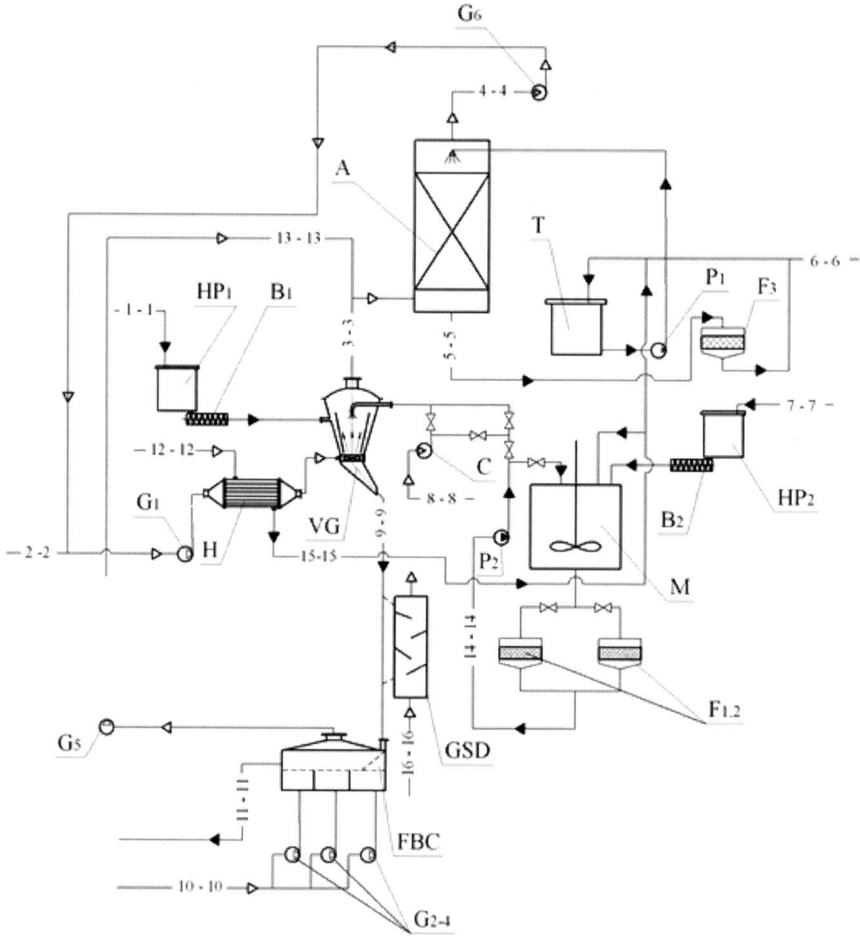

Elements of the installation: VG – a vortex granulator; H – a heater; GSD – a gravitational shelf dryer; FBC – a fluidized bed cooler; A – an absorber; F – a filter; M – a mixer; B – a batcher; HP – a hopper; G – a gas blower; P – a pump; T – a tank; C – a compressor.

The main flows: 1-1 – a seeding agent; 2-2 – manufacturing air; 3-3 – polluted air; 4-4 – purified air; 5-5 – polluted water; 6-6 – water; 7-7 – substandard granules; 8-8 – air for the spraying of the liquid materials (solution, melt); 9-9 – the product; 10-10 – air for cooling of granules; 11-11 – granules for packaging; 12-12 – steam; 13-13 – dusty gas; 14-14 – liquid materials (solution, melt); 15-15 – water condensate; 16-16 – a drying agent.

Figure 1. A unit for the production of PAN granules by using of a vortex granulator and/or a gravitational shelf dryer.

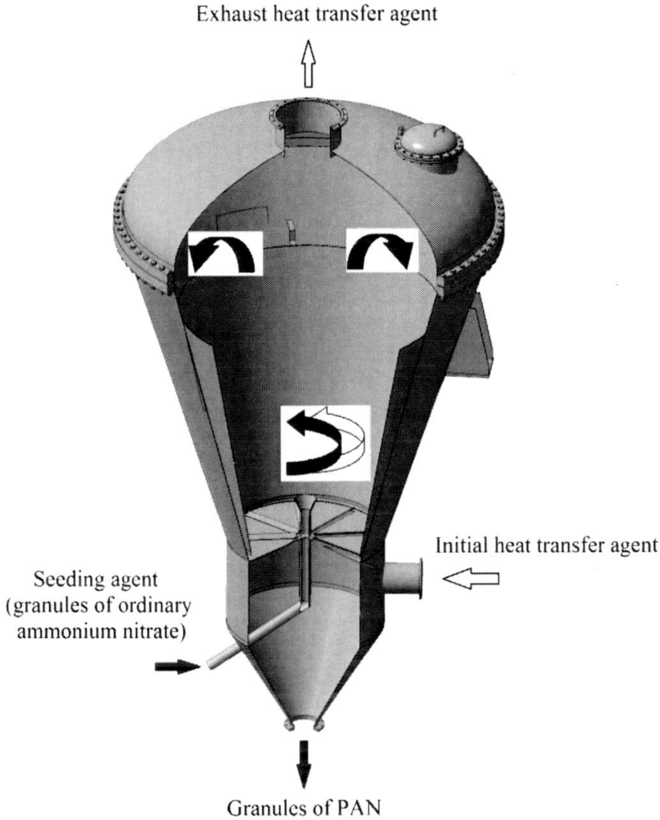

Figure 2. A vortex granulator.

Depending on the weight (size), the granules are classified into light and heavy (fine and large) fractions in response to the change of the circular and axial velocity of the axisymmetric vortex flows of the heating agent within the height of the work space. The produced porous granules are transferred from the upper overcut of the inner cone to the annular space toward the outlet sleeve. The heavy fraction remains in the work space, and while the weight is reducing, the fraction circulates in the inner cone and then moves to its upper overcut. The exhaust heating agent is removed from the granulator through the sleeve in its upper part.

The gravitational shelf dryer operates as follows (Figure 3).

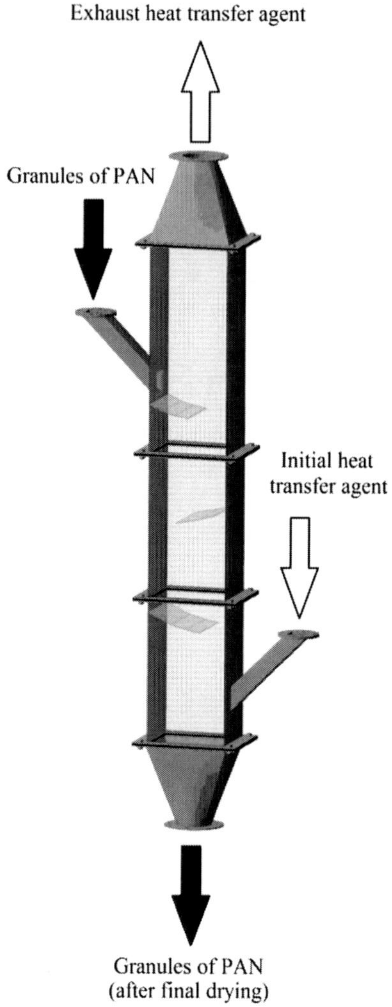

Figure 3. A gravitational shelf dryer.

The drying agent is supplied to the bottom of the unit, then it is uniformly distributed within the body overcut, and after that it is lifted. Thus, the agent goes through a certain amount (which is calculated) of the perforated tilted shelves (three of them in this case), followed by the removal from the upper part of the unit. While the drying agent is moving within the body, the dispersed material subjected to drying is supplied to the upper shelf. Under the counterflow impact of the drying agent on the upper tilted contact shelf, the dispersed material is heated, and fine fractions are removed in order to balance the roughness of the weighted layer. The length of the upper tilted contact shelf ensures the residence time of the dispersed material, which enables full heating of the material, the separation of the dispersed fines and the initial removal of the unbound moisture from the dispersed material's surface at the permanent velocity of the heating.

While drying by the dying agent streams, the dispersed material is gradually moving in the upper shelf, getting removed through the free edge of the shelf and is supplied to the middle shelf. The movement in the middle shelf is complemented by further intense removal of the unbound moisture from the dispersed material's surface at the permanent velocity of the heating and under the counterflow impact of the drying agent. The dispersed material is gradually removed through the free edge of the shelf, and then it is supplied to the lower (bottom) shelf. The movement of the dispersed material in the lower shelf is complemented by the removal of the bound moisture from the material depth at the reducing velocity of the heating. The dried dispersed material is removed from the unit.

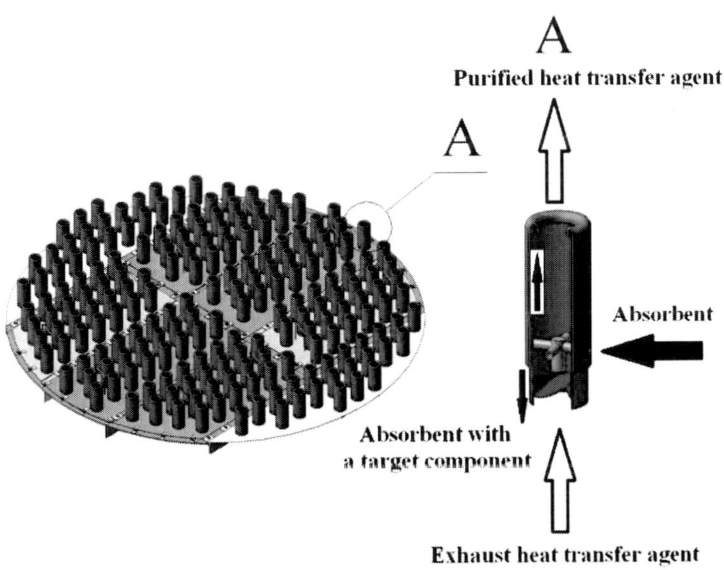

Figure 4. A vortex mass transfer and separation section.

Table 1. The fundamental algorithm for the calculation of the PAN production installation

A theoretical base	A unit (elements according to Figure 1)	The calculated value	A software product
The system of Reynolds Equation and the Continuity Equation	*VG*	The area of average velocities (vertical, circumferential, radial) of the gas flow in a working space with a height-variable cross section.	Conical channel©
The system of Navier-Stokes Equations and the Continuity Equation	*VG*	The area of the instantaneous velocity (vertical, circumferential, radial) of the gas flow in a working space with a height-variable cross section.	Vortex Flow©
The system of the equation of motion of granules, the kinetic equation of heating and drying of the capillary-porous body	*VG*	- the area of the instantaneous velocity (vertical, circumferential, radial) of granules in a working space with a height-variable cross section; - the motion trajectory of the granules; - the "hydrodynamic" residence time of the granules in the vortex granulator; - the area of the temperatures and moisture of the granules; - time for heating and drying of a granule; - the "thermodynamic" residence time of the granules in a vortex granulator.	Vortex Granulator©
A force analysis of the dispersed phase motion, the balance of the dispersed phase in a swirling gas flow.	*VG*	- the distribution of the granules of different sizes (weight) within the height of the working space of the vortex granulator; - the stable operating range of the vortex granulator; - the separation of the fine fractions and dust.	Classification in vortex flow©
A force analysis of the dispersed phase motion	*VG* *GSD*	The residence time of the granule in a sectioned working space of the unit.	Multistage fluidizer®
A force analysis of the continuous and dispersed phase motion	*VG* *A*	The trajectory of the gas flow and the liquid film motion in the mass transfer and separation element.	Vortex Tray©

The vortex mass transfer and separation section for the exhaust gas purification operates as follows (Figure 4).

A rotating gas-liquid layer is formed inside the stationary contact pipe, where the gas is supplied through a swirler at the bottom of the pipe, and the liquid flow is supplied from the tray's surface through the liquid feeding pipe. The swirled gas flow disperses the liquid under the swirler (thus, the developed surface of the contact phases is formed). The gas-liquid flow further gets split under the influence of the centrifugal forces. The liquid is pushed to the periphery and in a form of a film goes up in the inner walls of the contact pipe toward the film-remover under the aerodynamic forces; the separated gas is removed through the upper part of the contact pipe. The film of the liquid goes through the film-remover, then along the gap between the contact pipe and the film-remover and flows down through the drain hole on the tray surface.

The software products are based on the original mathematical models of the hydrodynamics of the movement of single and multiphase vortex flows, the heating and drying kinetics [6, 18-30]. Due to the limited space of this article, the reader may learn about these software products from original sources, and this work presents the calculation for the PAN production installation in a form of the fundamental algorithm (Table 1).

All the presented calculation parameters are further used in the methodology of the engineering calculations for the granulation devices.

RESULTS AND DISCUSSION

Conical Channel©

The software programme Conical channel© is applied to calculate the axisymmetric gas flows in the channels, including the swirling flows in the diffusers.

The output data for the calculations are the geometrical configuration of the channel, the gas properties and the search parameters in the inlet overcut. The calculations are used to determine the velocity and pressure fields within the whole range of the measurement area, and as a result to determine the value of the energy loss between the inlet and outlet overcuts of the channel.

The programme displays the calculated data in a form of the dependency graph. The programme starts with the opening or creation of the new configuration file of the measurement area. Figure 5 shows the example of the measurement area configuration.

```
[Left_Bound]
Nodes=12
Point_1=0.00 0.090
Point_2=0.00 0.08999 2.53 0.02
Point_3=0.00 0.085 2.53 0.02
Point_4=0.00 0.080 2.72 0.00
Point_5=0.00 0.075 2.81 0.00
Point_6=0.00 0.070 2.84 –0.05
Point_7=0.00 0.065 2.73 –0.10
Point_8=0.00 0.060 2.62 –0.05
Point_9=0.00 0.055 2.73 0.10
Point_10=0.00 0.050 2.98 0.37
Point_11=0.00 0.04401 2.98 0.00
Point_12=0.00 0.044
```

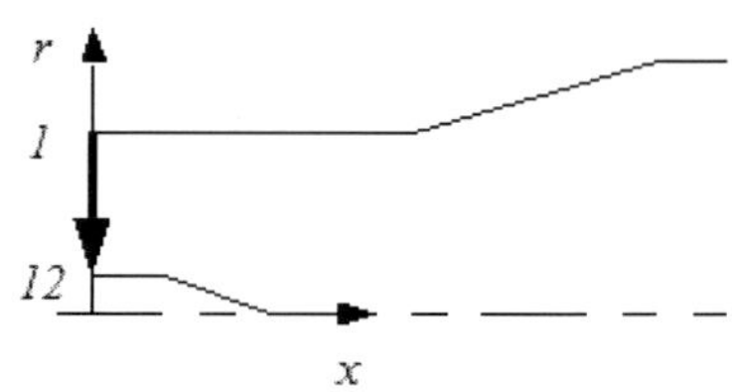


```
[Right_Bound]
Nodes=2
Point_1=0.550 0.118
Point_2=0.550 0.00

[Bottom_Bound]
Nodes=4
Point_1=0.00 0.044
Point_2=0.070 0.011
Point_3=0.075 0.00
Point_4=0.550 0.00

[Top_Bound]
Nodes=4
Point_1=0.00 0.090
Point_2=0.200 0.090
Point_3=0.330 0.118
Point_4=0.550 0.118
```

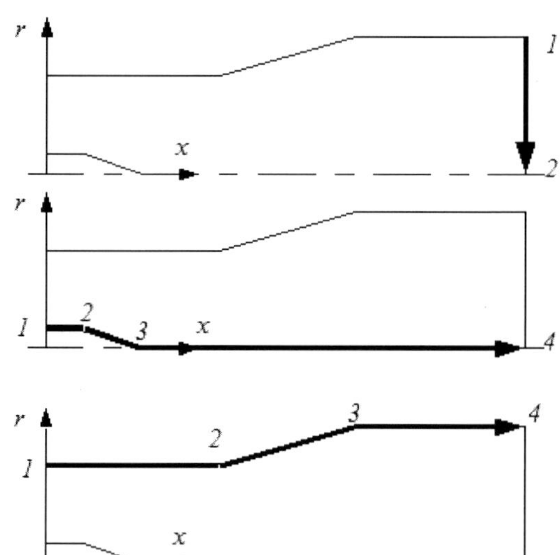

Figure 5. An example of the measurement area configuration.

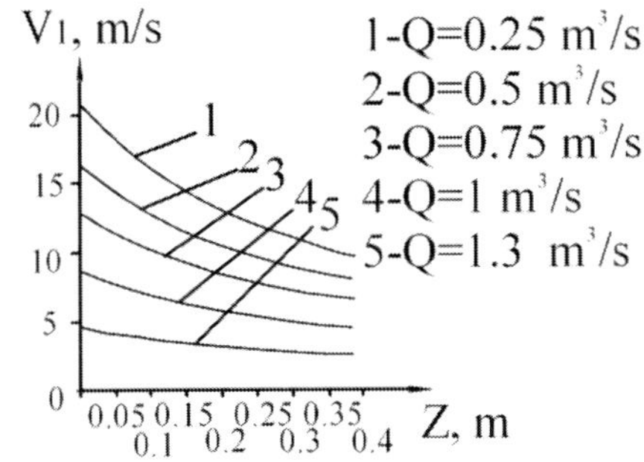

Figure 6. The calculated change of the longitudinal velocity of the gas flow.

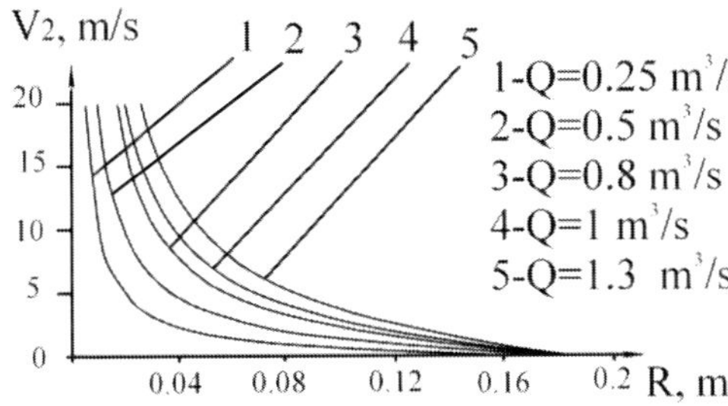

Figure 7. The calculated change of the radial velocity of the gas flow.

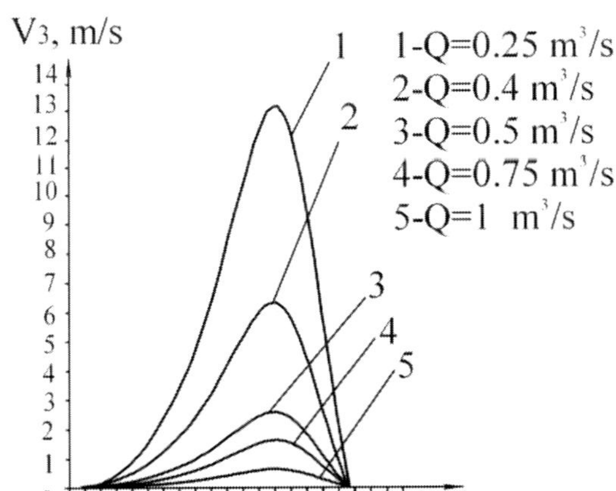

Figure 8. The calculated change of the circumferential velocity of the gas flow.

After selecting the necessary parameter, the programme builds the calcultaion grid. The analysis of the components of the full movement velocity requires determination of the character of the dependencies and their comparison within one comparative graph (as depicted in Figures 6-8).

Vortex Flow©

The software programme Vortex Flow© is applied to calculate the hydrodynamic performance of the swirling gas flow and particles in it. The programme may be used for hydrodynamic calculations of the flows in the vortex heat-mass transfer (a granulator, dryers) and hydromechanical (classifiers, cyclones) equipment with the variable cross-section of the working chamber. The result of the calculations is the value of the velocity components of the movement of gas flows and particles in the work space. The programme displays the calculated data (or prints if necessary) in a form of the listed results or a dependency graph.

The programme was written in Delphi and runs on the OS Windows.

At the first stage a user selects the calculated data in an array (Figure 9). After the calculations completed, an array of the numeric values of the specified parameter is displayed in each cell of the calculation grid according to their number in the output data.

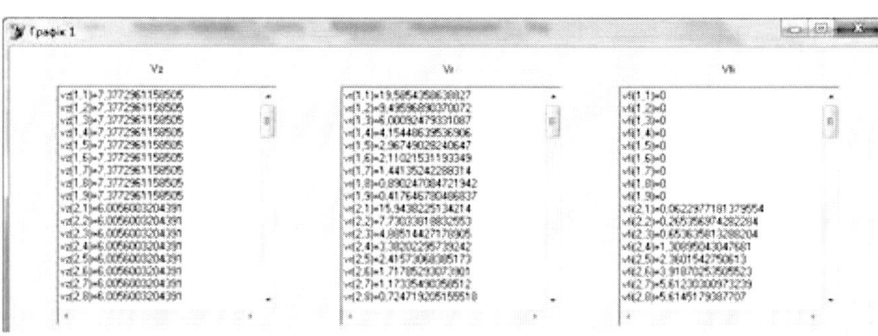

Figure 9. An array of the calculated data (an example of the calculations of the gas flow velocity).

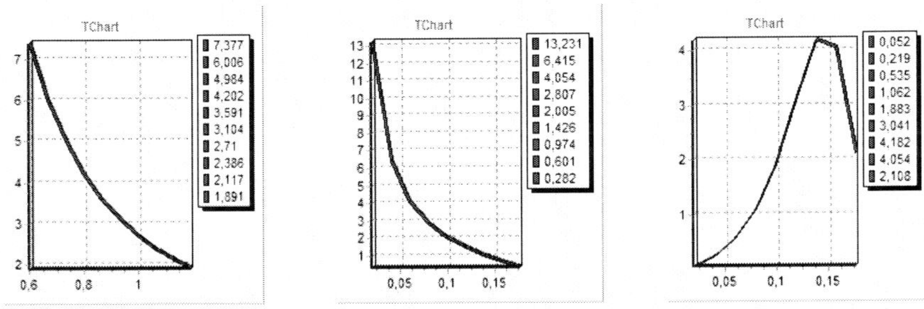

Figure 10. Display of the graphical dependences (an example of the calculations of the gas flow velocity).

After correct entering of the additional data, a dependency graph is displayed (Figure 10).

Thus, the display window for the calculation results generally looks as shown in Figure 11.

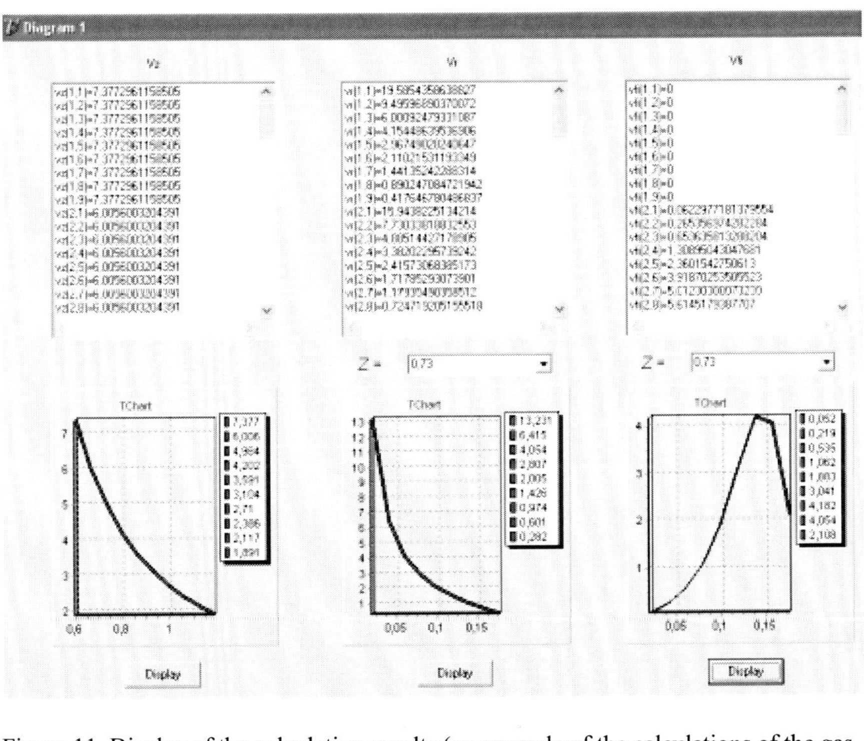

Figure 11. Display of the calculation results (an example of the calculations of the gas flow velocity).

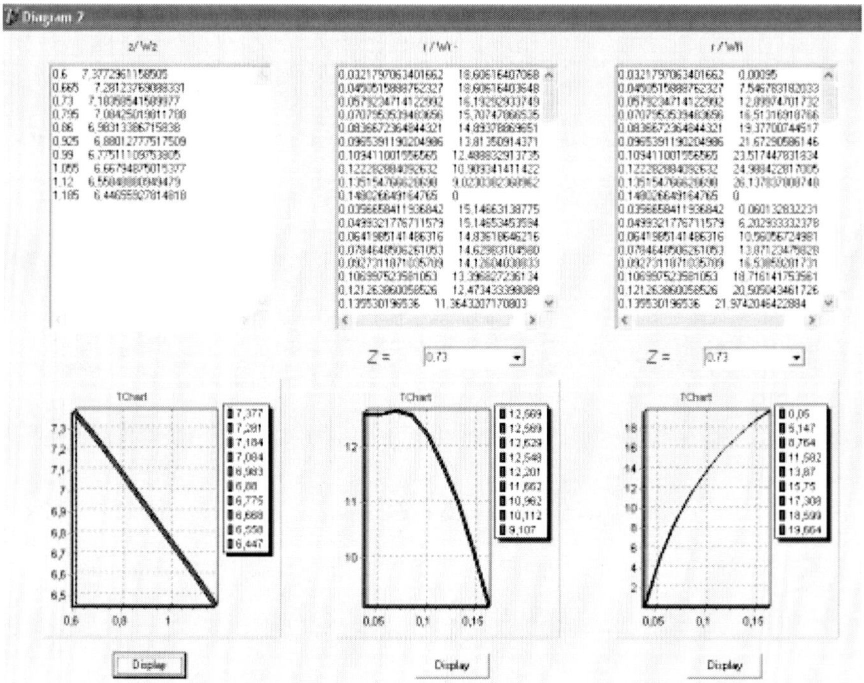

Figure 12. Display of the calculation results (an example of the calculations of the particle velocity).

The similar procedure is applied to display the calculations of the particle movement velocity (Figure 12).

A user can calculate the velocity of the gas flow and a particle separately, and to conduct consecutive calculations of these parameters by pressing a relevant button and moving to the required dialog box.

Vortex Granulator©

The programme Vortex Granulator© uses 9 classes and one file of styles.

The Main.java class enables display of the main menu and connection with other files of the programme.

The ControllerInput.java class provides entering of the data for calculations of the velocity components of the gas flow and granules and total velocities of these phases.

The Theory.java class provides a theoretical report on the vortex granulator, its advantages and disadvantages.

The SpeedGas.java and Granuly.java classes depict a dependency graph of the velocity components (vertical, radial and circular) of the gas flow and granules.

The Traektory.java class depicts thc total velocities.

The Functional.java class stores input values of the programme and contains methods of solving and formation of the data arrays for the dependency graph.

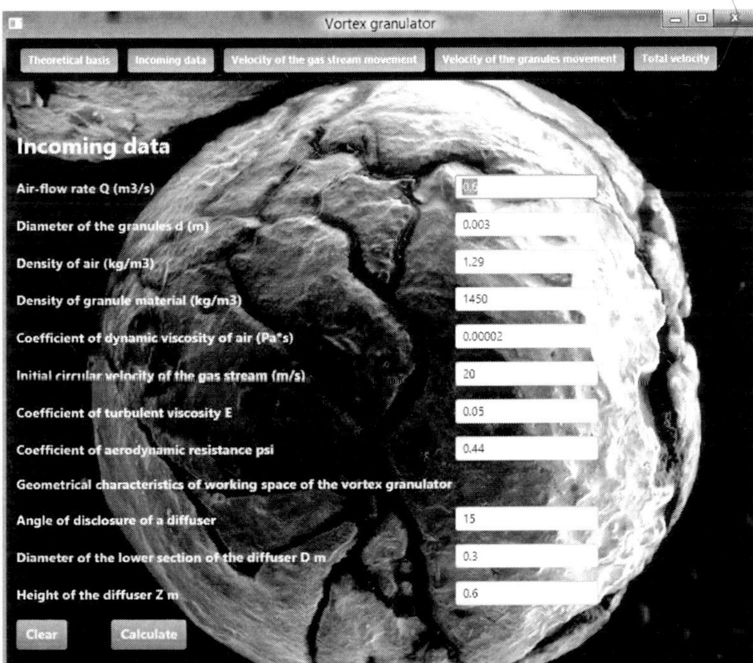

Figure 13. An incoming data window.

The Error.java and AlertBox.java classes are devoted to check the validation of the entered data and to handle any errors while running the programme.

The main class of Vortex Granulator© is Main.java, which displays the main menu and gets other files connected to it. After the programme started, the data entering class is the first to be connected by default. When the incoming parameters (Figure 13) are entered, the programme launches their validation with Error.java class. If errors of the entering occur, the programme runs the AlertBox.java class, responsible for the information about the error occurred. The validated data are further stored and calculated in the Functional. java class.

The programme contains 3 calculation areas: gas flow velocity components, granule velocity components, total velocities of the gas flow and granules. When the menu items are activated, the programme calculates radial, vertical and circular velocity components of the gas flow (Figure 14) and granules (Figure 15).

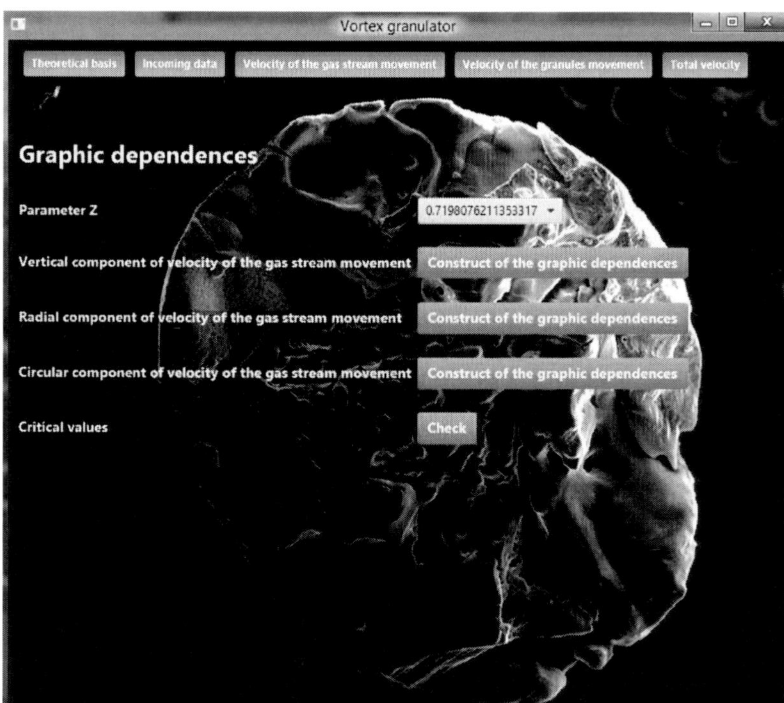

Figure 14. A window of the graphical dependences of the gas flow velocity components.

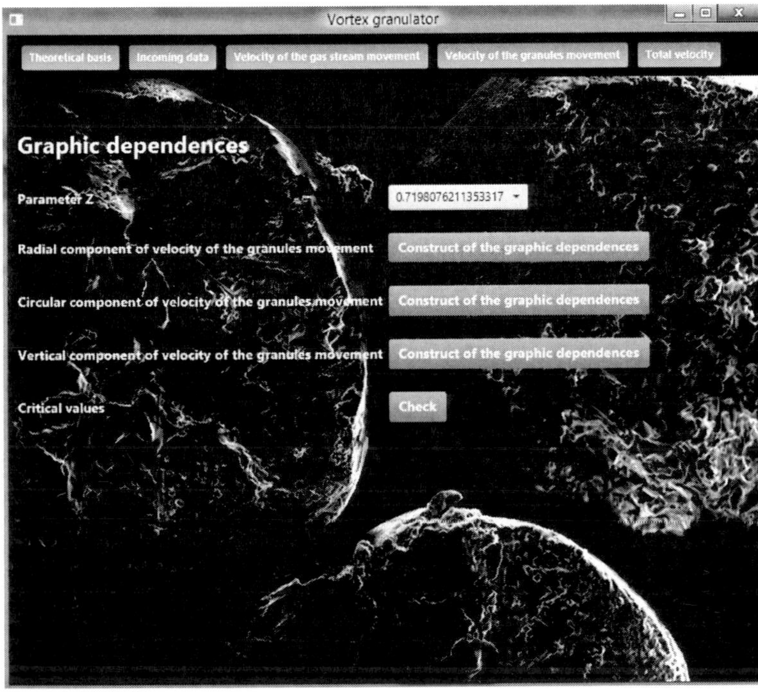

Figure 15. A window of the graphical dependences of the granule's velocity components.

The calculation results may be displayed in a form of the dependency graph.

The main methods of the calculation of the components and total velocities of the gas flow and granules are as follows:

- public speedGas – calculates a vertical component of the gas flow velocity given the height of the work space of the granulator (Figure 16);
- public Vr – calculates a radial component of the gas flow velocity given the radius of the vortex granulator (Figure 17);
- public Vfi - calculates a circular component of the gas flow velocity given the radius of the vortex granulator (Figure 18);
- public RungeKutt – applies the Runge-Kutta Method to determine the radial component of the granules velocity;

- public Wr – calculates a radial component of the granules velocity given the radius of the vortex granulator (Figure 19);
- public Wfi – calculates a circular component of the granules velocity given the radius of the vortex granulator (Figure 20);
- public RungeKuttWz - applies the Runge-Kutta Method to determine the vertical component of the granules velocity;
- public Wz – calculates a vertical component of the granules velocity given the height of the work space of the granulator (Figure 21);
- public resultSummSpeed – calculates the total velocity of the gas flow given the radius of a vortex granulator (Figure 22);
- public resultSummGranul – calculates the total velocity of granules given the radius of a vortex granulator (Figure 23).

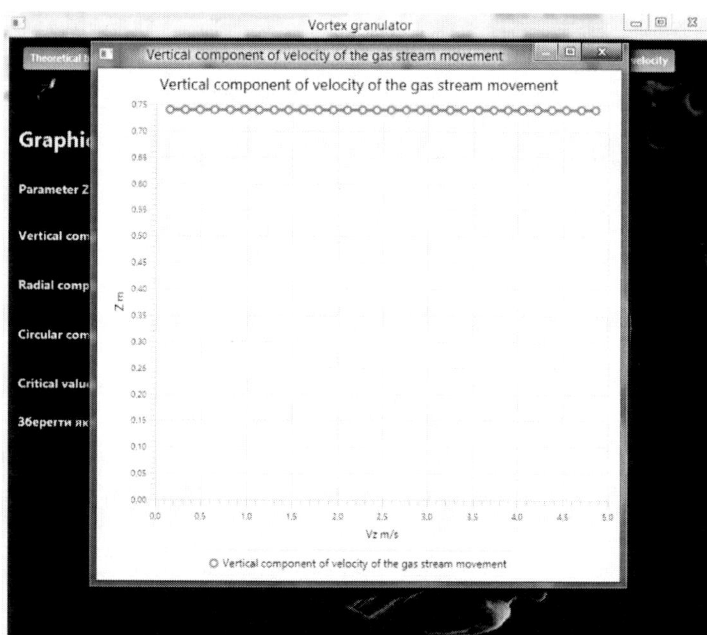

Figure 16. A vertical component of the gas flow velocity.

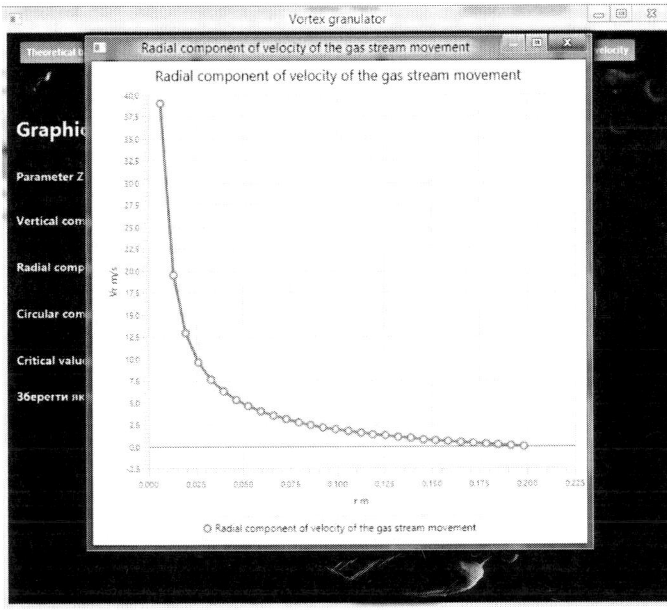

Figure 17. A radial component of the gas flow velocity.

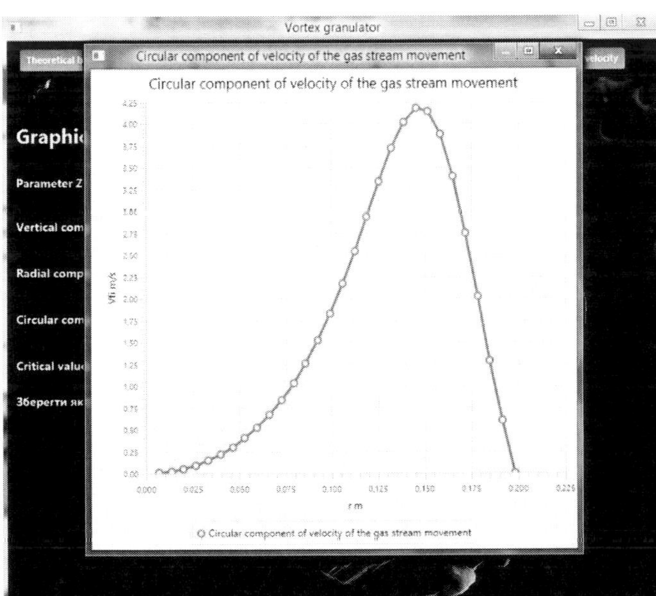

Figure 18. A circular component of the gas flow velocity.

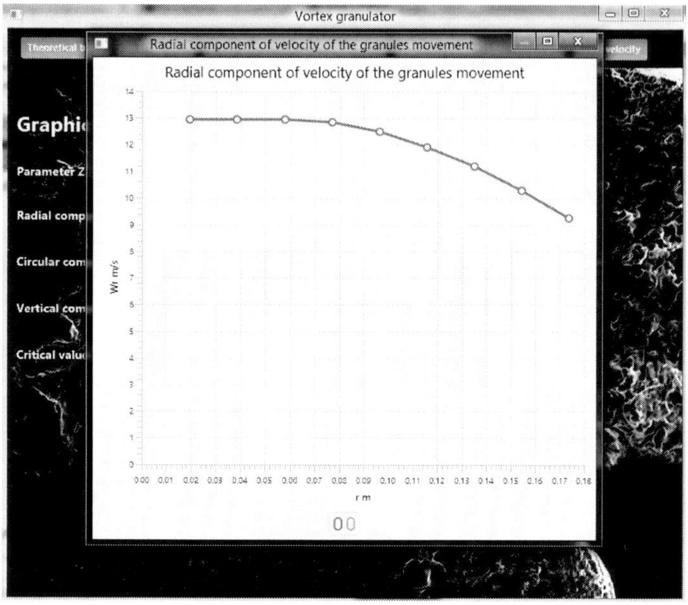

Figure 19. A radial component of the granule's velocity.

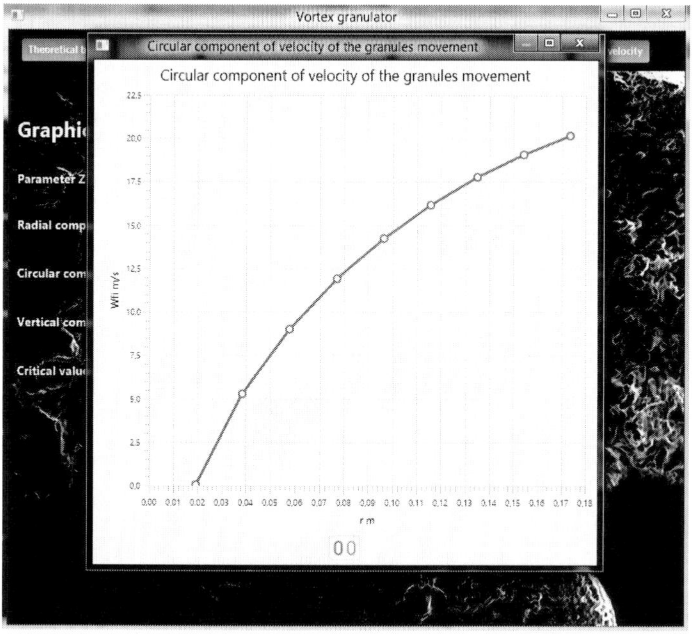

Figure 20. A circular component of the granule's velocity.

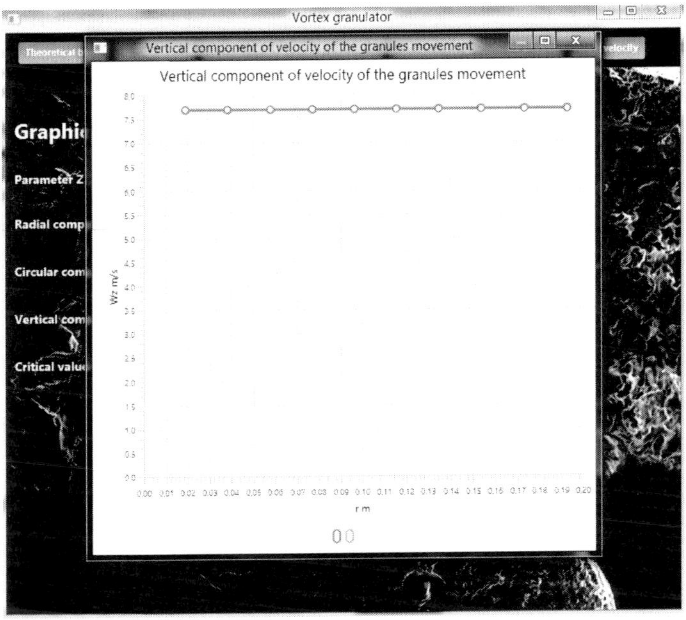

Figure 21. A vertical component of the granule's velocity.

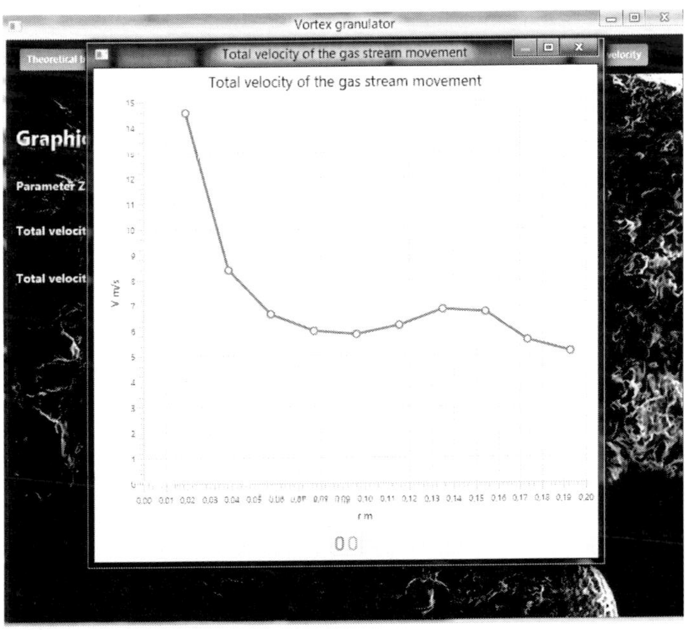

Figure 22. Total velocity of the gas flow.

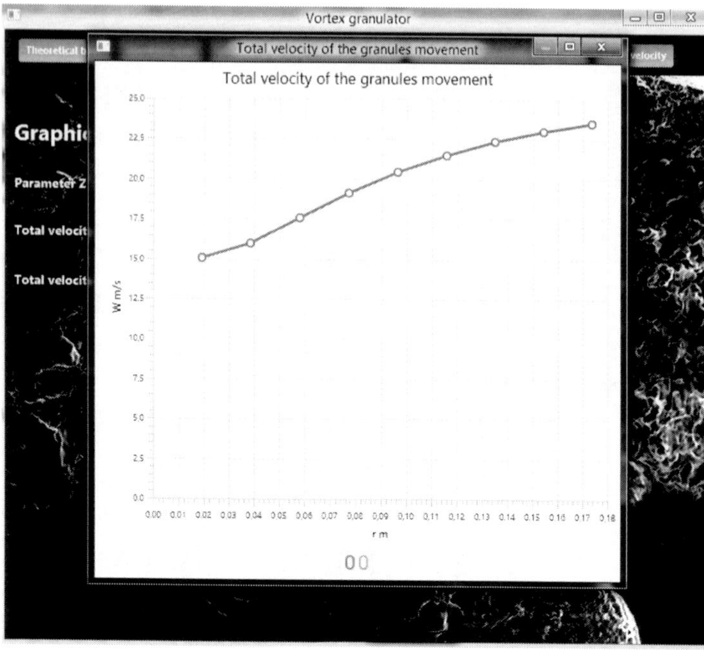

Figure 23. Total velocity of the granules.

Classification in Vortex Flow©

The programme Classification in vortex flow© contains 7 classes and 1 file of styles.

The Main.java class enables display of the main menu and connection with other files of the programme.

The DataInput.java class stores incoming data of the programme and contains methods of solving and formation of the data arrays for the dependency graph.

The ControllerInputSpeedGas.java class contains a class for the graphical interface display.

The Graph.java class displays the obtained results in a form of the dependency graph.

The Excel generator.java class creates a document as an electronic table with obtained results.

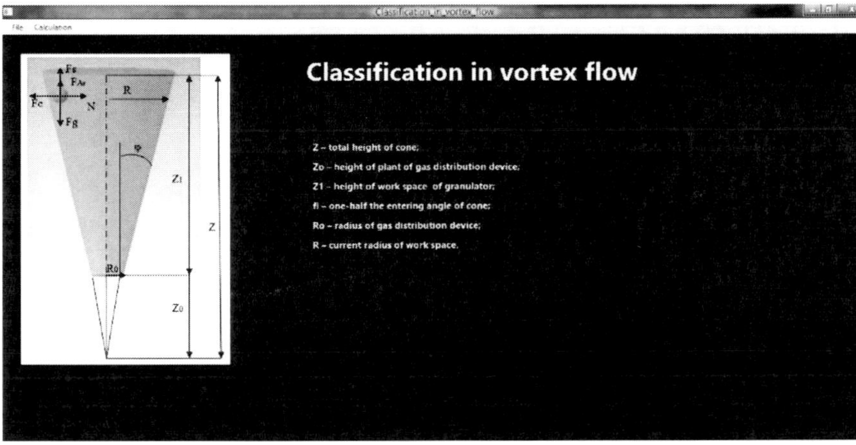

Figure 24. A calculation area of the granulator.

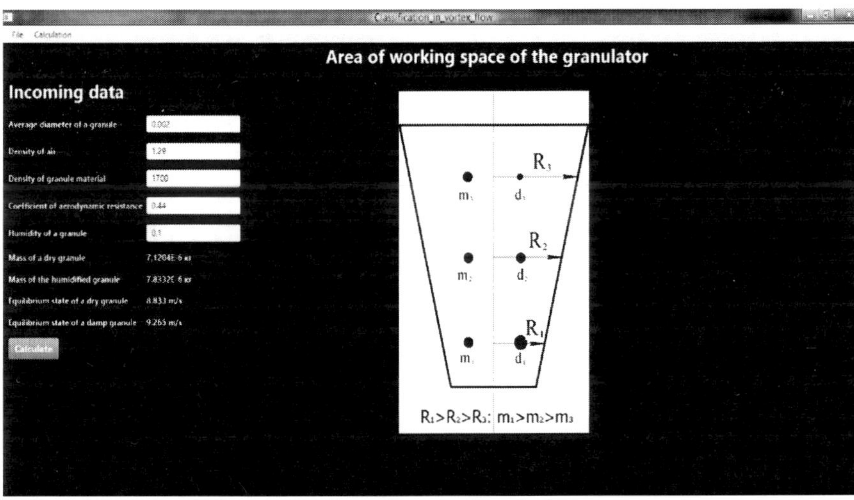

Figure 25. Initial data entering. Gas flow velocity.

The Error.java and AlertBox.java classes are devoted to check the validation of the entered data and to handle any errors while running the programme.

The main class of Classification in vortex flow© is Main.java, which displays the main menu and gets other files connected to it. After launching, the programme displays the model (Figure 24) and general parameters of the granulator (Figure 25).

The programme contains 3 calculation areas:

1. The gas flow velocity (Figure 25).
2. The geometry of the work space:
   - calculations in case of the variable consumption of air – a display of dependencies of the height of the work space and the radius of the cross-section on the air consumption for damp and dry granules (Figure 26).
   - calculations in case of a variable opening angle of the diffuser –a display of dependencies of the height of the work space on the angle of the diffuser opening for damp and dry granules (Figure 27).
   - calculations of the granules distribution by fractions in the work space of the granulator – display of the distribution of the granules with different diameters within the height of the vortex granulator's work space (Figure 28).

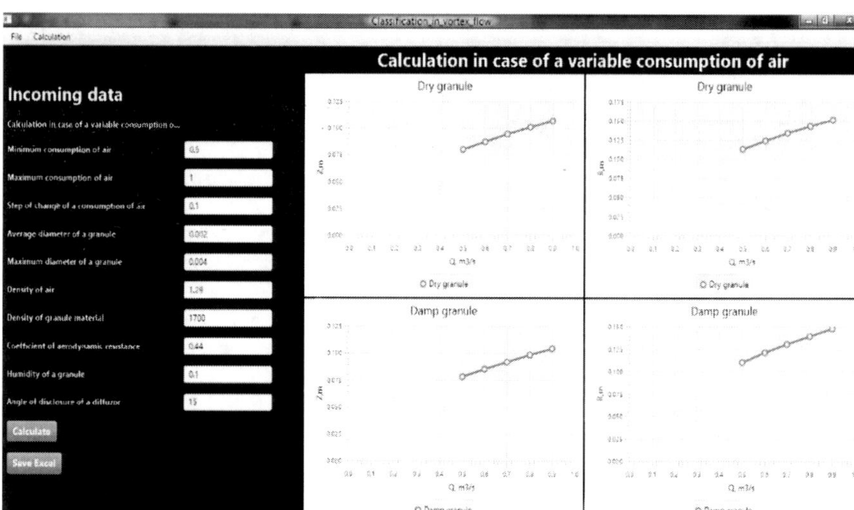

Figure 26. Calculations of the geometry of the work space of the vortex granulator in case of the variable consumption of air.

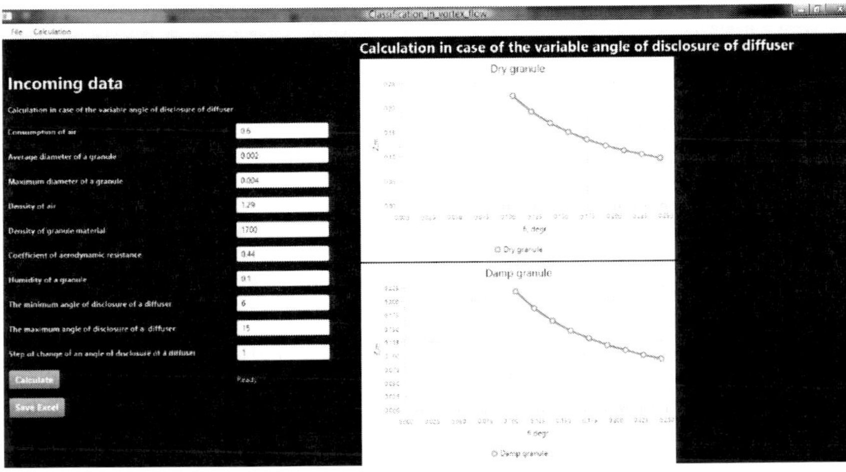

Figure 27. Calculations of the geometry of the work space of the vortex granulator in case of the variable angle of opening of the diffuser.

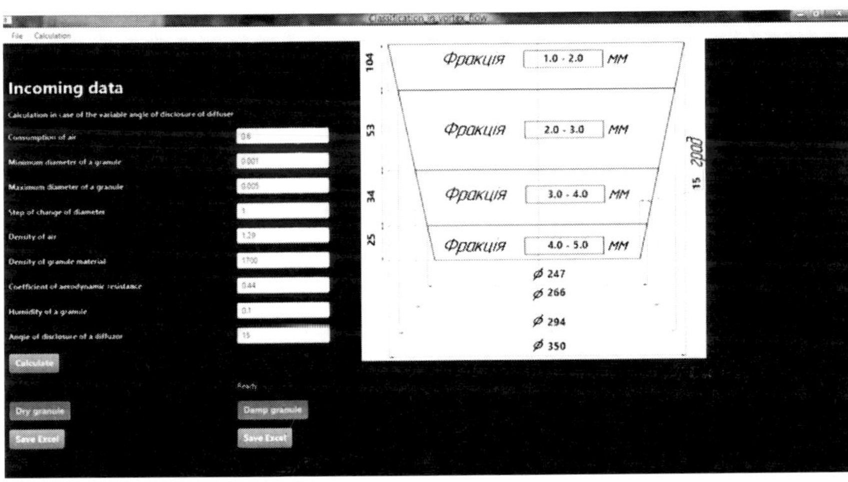

Figure 28. Calculations of the distribution of granules by fractions in the working space of the granulator.

3. Calculations of the mass and kinetics of granule heating:
   - calculations of the granule's mass (Figure 29);
   - calculations of the kinetics of the granule heating (a distribution of the temperature within the granule's radius) (Figure 30);

- calculations of the kinetics of the granule heating (a time change of the granule's temperature) (Figure 31).

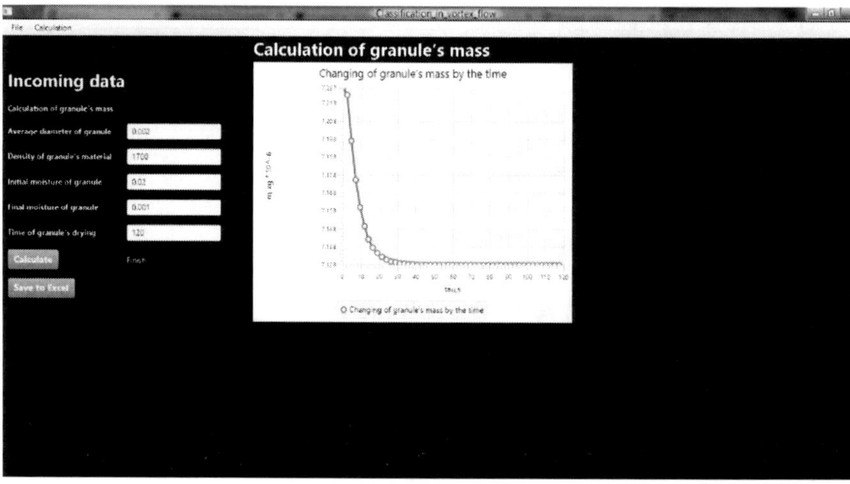

Figure 29. Calculations of granule's mass.

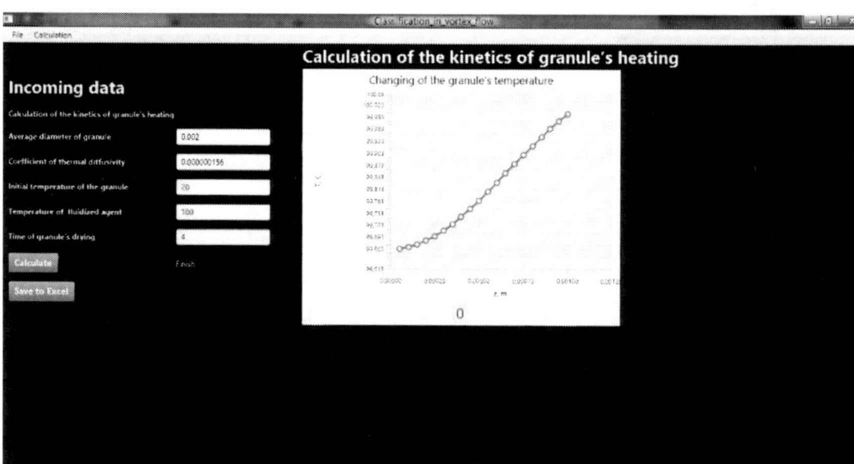

Figure 30. Calculations of the kinetics of granule heating (the distribution of the temperature within the granule's radius).

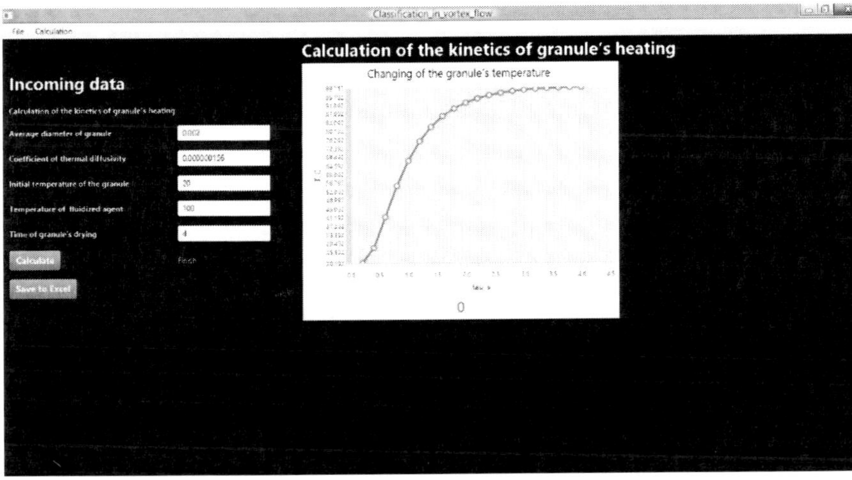

Figure 31. Calculations of the kinetics of the granule heating (time change of the granule's temperature).

To calculate the necessary values, a user selects the relevant item, enters data and presses "Calculate." The programme will validate the data, calculate the relevant values and display the obtained data in a form of a dependency graph. The programme Classification in vortex flow© is also able to save the obtained result in a form of the e-table.

The main methods of the calculations are:

- public double masSuchGranuly – calculates the dry granule's mass;
- public double masVologGranuly – calculates the damp granule's mass;
- public double speedSuchGranuly – calculates the dry granule's velocity;
- public double speedVologGranuly – calculates the damp granule's velocity;
- public double heightVitanSuchGranuly – calculates the flight height of the dry granule;

- public double heightVitanVologGranuly – calculates the flight height of the damp granule;
- public double radiusBottomPererizSuchGranuly – calculates the radius of the lower overcut of the granulator for a dry granule;
- public double radiusBottomPererizVologGranuly – calculates the radius of the lower overcut of the granulator for a damp granule;
- public double simpsonLanst – applies Simpson's method and calculates integrals;
- public double sumTemperatur – calculates the kinetics of the granule heating (a distribution of the temperature within the granule's radius)
- public double sumTemperaturtau – calculates the kinetics of the granule heating (a time change of the granule's temperature).

Multistage Fluidizer®

The programme Multistage fluidizer® uses HTML HyperText Markup Language, CSS Cascading Style Sheets and JavaScript programming language (including a JQuery library). HTML is a markup of the web apps, CSS designs pages, and JavaScript calculates and transfers data, creates animation and validates data. The data are checked in the validation box of the Java Script. The input info box accepts and further presents incoming data values. The calculation box presents calculations according to formulas.

Index.html (Figure 32) is the main page of the web app. It displays the main menu, the main calculation of the gas flow and a jump to next pages (an example of the page is shown in Figure 33), where the main dependences between key characteristics of the calculation of the gas flow and the residence time of the material on the shelf are calculated and graphical dependences are built.

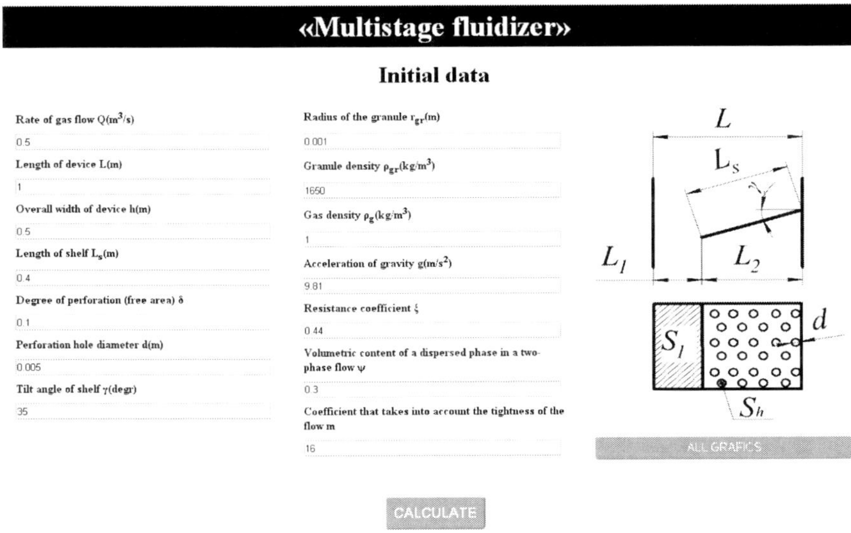

Figure 32. The main page of the programme.

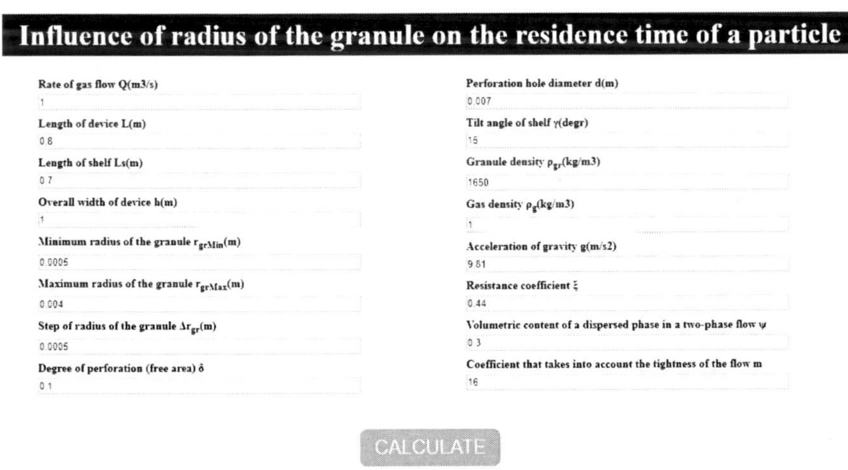

Figure 33. A page with calculations on how various parameters influence the residence time of the particles in the device.

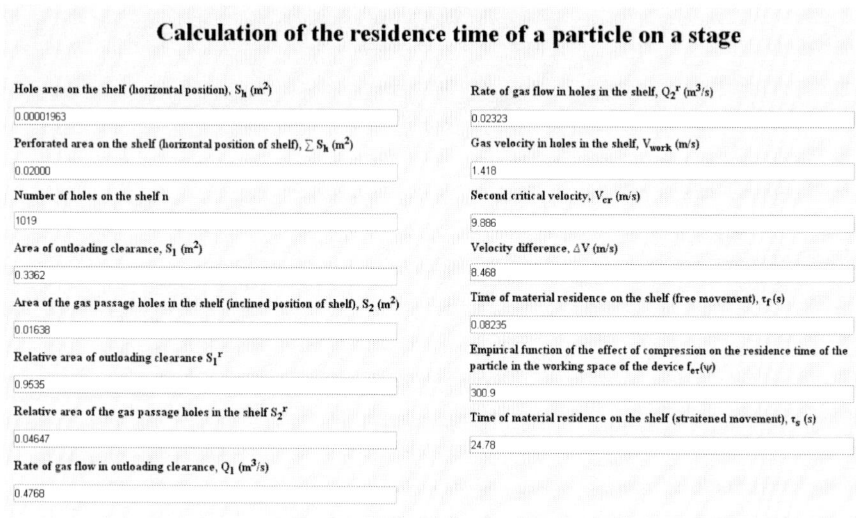

Calculation of the residence time of a particle on a stage

Hole area on the shelf (horizontal position), S_h (m^2)

0.00001963

Perforated area on the shelf (horizontal position of shelf), $\sum S_h$ (m^2)

0.02000

Number of holes on the shelf n

1019

Area of outloading clearance, S_1 (m^2)

0.3362

Area of the gas passage holes in the shelf (inclined position of shelf), S_2 (m^2)

0.01638

Relative area of outloading clearance S_1^r

0.9535

Relative area of the gas passage holes in the shelf S_2^r

0.04647

Rate of gas flow in outloading clearance, Q_1 (m^3/s)

0.4768

Rate of gas flow in holes in the shelf, Q_2^r (m^3/s)

0.02323

Gas velocity in holes in the shelf, V_{work} (m/s)

1.418

Second critical velocity, V_{cr} (m/s)

9.886

Velocity difference, ΔV (m/s)

8.468

Time of material residence on the shelf (free movement), τ_f (s)

0.08235

Empirical function of the effect of compression on the residence time of the particle in the working space of the device $f_{\tau\tau}(\psi)$

300.9

Time of material residence on the shelf (straitened movement), τ_s (s)

24.78

Figure 34. Calculation results.

After the data are entered, they are validated, and if correct, the button CALCULATE will show the results in a form of a table (Figure 34).

The change of l_f and l_s values will show animation if one presses the button Show calculation (Figure 35 shows calculation of the length of a particle's track on the shelf during a specified time span). The functions clicker_lf and clicker_ls are used to create τ_f and τ_s animations correspondingly.

In JavaScript the calculations are performed with: JQuery libraries and table2excel, recorder objects, .val(), .append() methods to read and insert values into the fields, .removeClass() and .addClass() methods to delete and add classes, and .animate() method, which creates animation effects on any numeric CSS property.

Some graphical dependences are depicted in Figure 36. The programme allows obtaining of two- and three-dimensional dependency graphs.

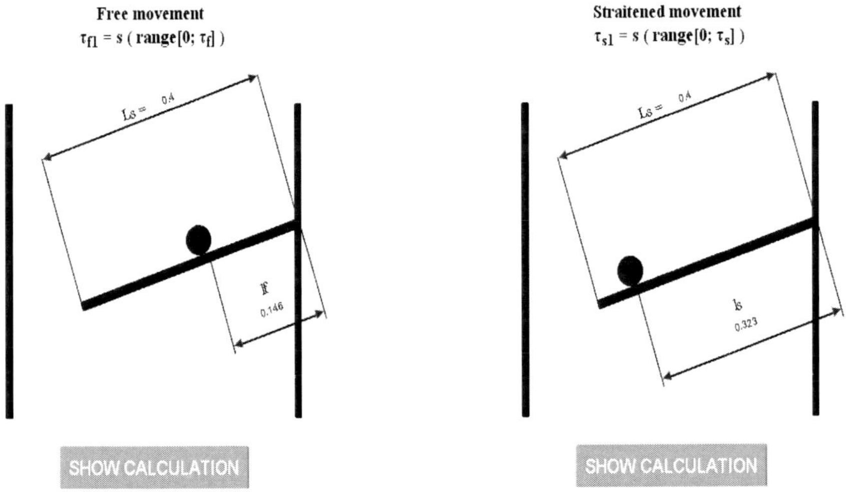

Figure 35. Calculations of the length of a particle's track on the shelf during a specified time span.

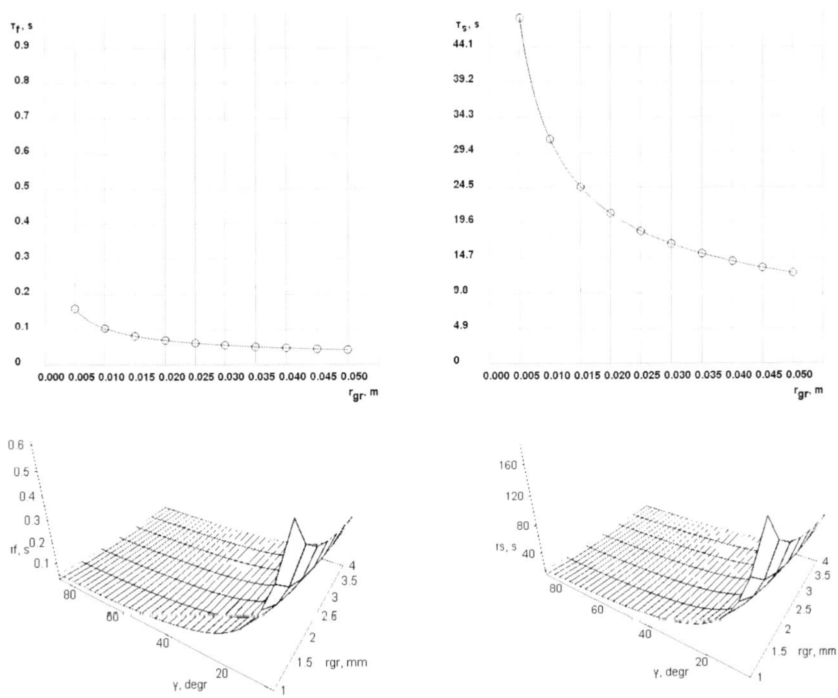

Figure 36. Examples of the calculation results.

Vortex Tray©

The programme Vortex tray© (Figure 37) determines the influence of the technological operation mode of the granulator on the size of the heat-mass transfer and separation element.

The computer modeling results under various conditions are shown in Figures 38-42.

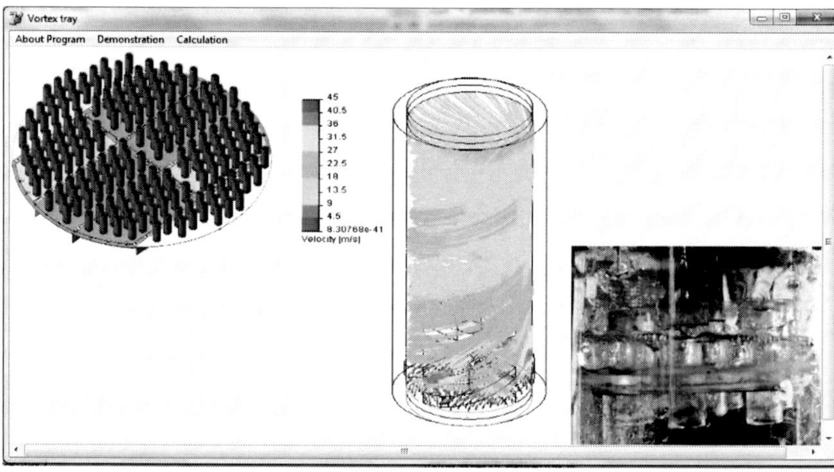

Figure 37. The Vortex tray© interface.

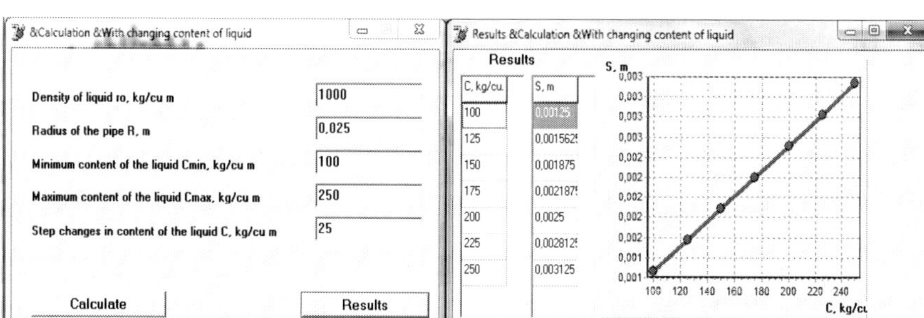

Figure 38. The thickness of the liquid film with the changing value of C.

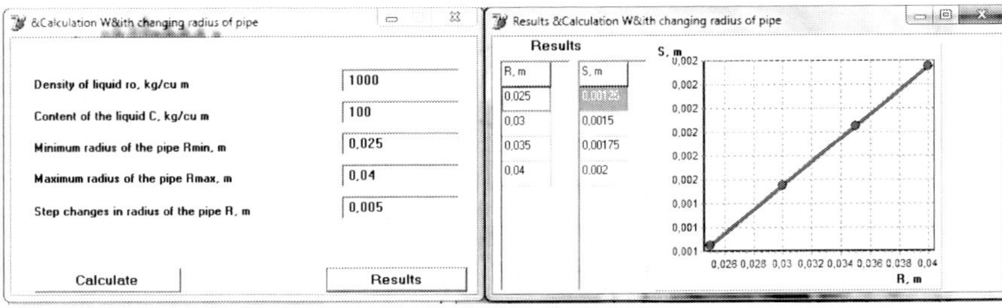

Figure 39. The thickness of the liquid film with the changing value of R.

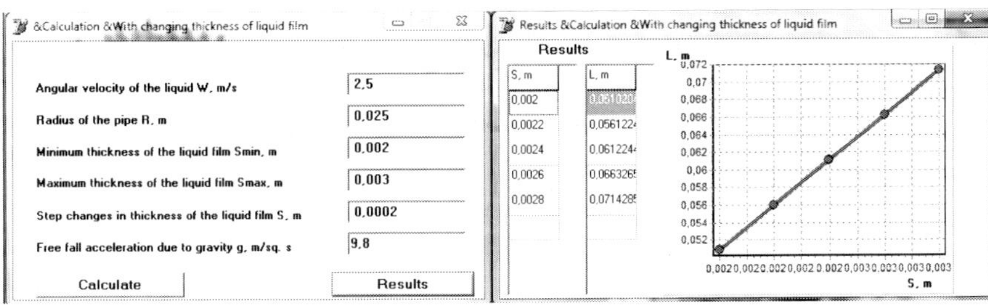

Figure 40. The height of the contact pipe with the changing value of *S*.

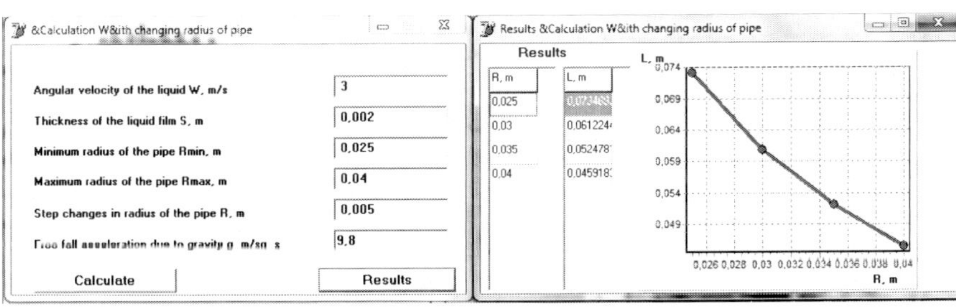

Figure 41. The height of the contact pipe with the changing value of *R*.

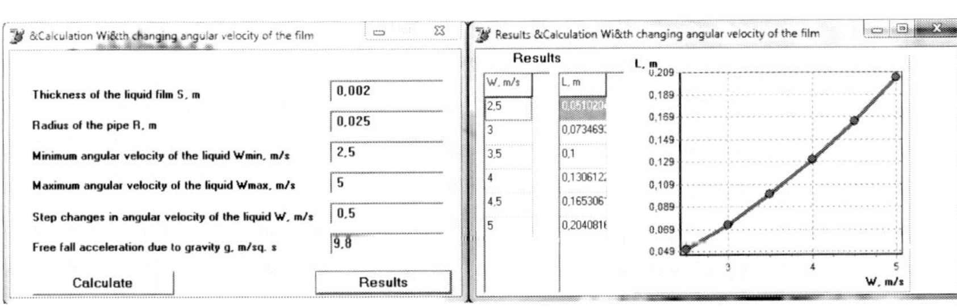

Figure 42. The height of the contact pipe with the changing value of W_{φ}.

Conclusion

Granulation Unit©

The Granulation Unit© cluster unites all the above-mentioned software products as separated structural blocks (Figure 43).

The proposed complex algorithm and the above-presented software products allow users to perform the optimization calculation for the main technological elements of the granulation unit at a stage of its design avoiding an expensive multifactor experiment.

The peculiarities of the software products also enable the readjustment of the equipment to the changes in the characteristics of raw materials and the productivity of the unit.

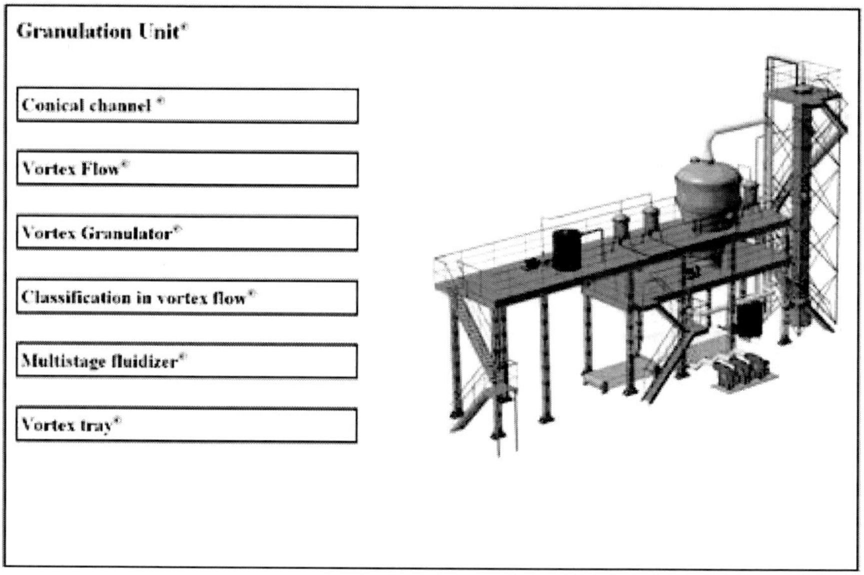

Figure 43. The Granulation Unit© cluster.

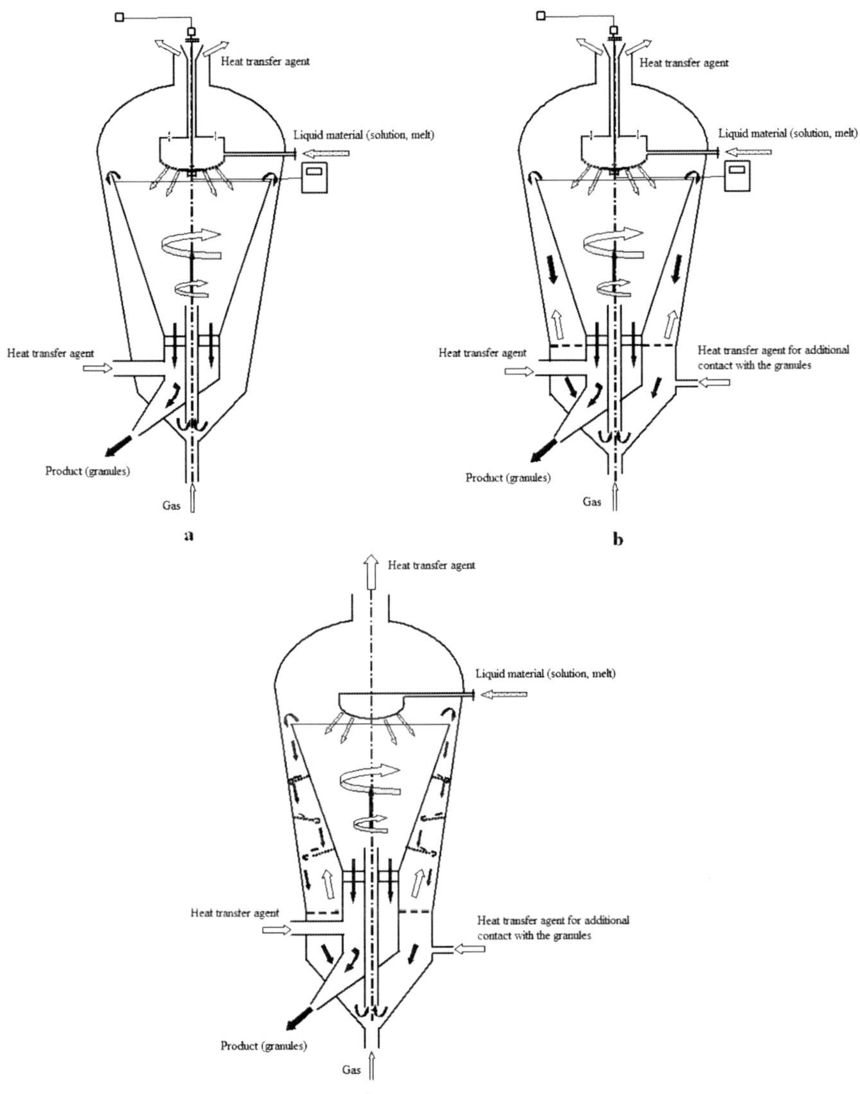

Figure 44. Vortex granulators: a – with vibrating spraying of melt; b – with the two-stage contact of granules and a heat transfer agent; c – with the multi-zone secondary contact of granules and a heat transfer agent.

The calculation results obtained with the software products and the cluster help to ensure, for example, the following:

- the minimal "hydrodynamic" residence time of the particles in the device, which won't exceed "thermodynamic" time (the minimal time of the dehydration of particles to a standard indictor; it is determined by the laws of the dehydration kinetics and thermodynamic indicators of the dehydration);
- the use of various temperature and moisture potentials of the heating agent in a separate part of the unit;
- variations of the unit construction;
- a possibility to apply recycling of the heating agent.

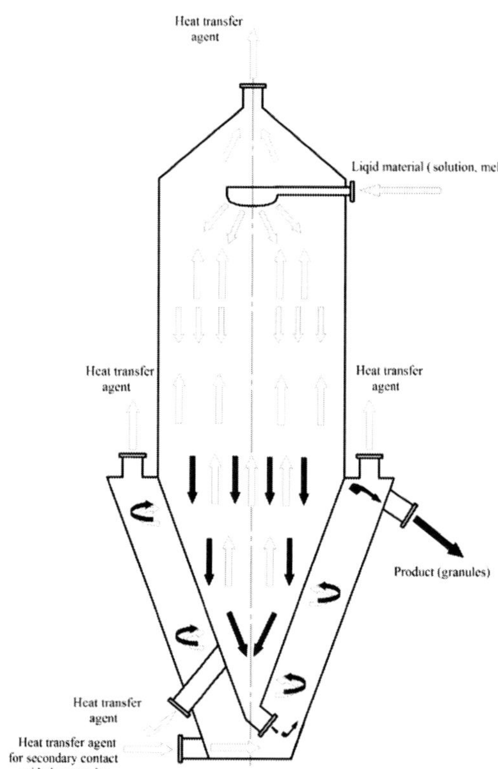

Figure 45. A dual-zone vortex granulator.

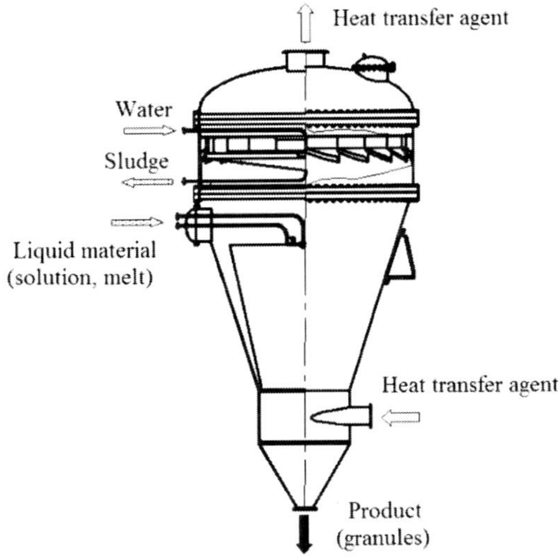

Figure 46. A vortex granulator with a separating device for gases cleaning.

Based on the technological and constructive calculations (performed with the above-described software products and the cluster), the design of the industrial model of the vortex granulator is presented (Figures 44-46).

Further investigations aim to develop an algorithm for exporting the constructive calculation results, obtained for the main technological facilities of the granulation unit, into the solid modeling programmes with further modeling of the devices and preparation of the engineering documentation of their production.

ACKNOWLEDGMENTS

This work was carried out under the project "Improving the efficiency of granulators and dryers with active hydrodynamic regimes for obtaining, modification and encapsulation of fertilizers", state registration No. 0116U006812. While preparing the article, the information about the original designs of vortex granulators (patents Nos 110403, 110992, 111406, 111592, 112021, 112293, 114521, Ukraine) and the author`s software products (certificates of authorship Nos 52659, 55360, 57043,

62692, 65140, 67472, Ukraine) was used. The authors thank the researchers of Processes and Equipment of Chemical and Refining Industries Department, Sumy State University (Ukraine) for their valuable comments during the preparation of the article.

References

[1] Dvornikov, N. A., Belousov, P. P. (2011). Investigation of a fluidized bed in a vortex chamber. *Journal of Applied Mechanics and Technical Physics*, 52(2), 206-211.

[2] Khalatov, A. A. (2010). Heat transfer and hydrodynamics in the fields of mass forces. *Journal of Engineering Physics and Thermophysics*, 83(4), 794-808.

[3] Kaewklum, R., Kuprianov, V. I. (2010). Experimental studies on a novel swirling fluidized-bed combustor using an annular spiral air distributor. *Fuel*, 89, 43-52.

[4] Barsky, E., Barsky, M. (2006). *Cascade Separation of Powders*. Cambridge: Cambridge Int Science Publishing.

[5] Ashcraft, R. W., Heynderickx, G. J., Marin, G. B. (2012). Modeling fast biomass pyrolysis in a gas-solid vortex reactor. *Chemical Engineering Journal*, 207-208, 195-208.

[6] Artyukhov, A. E. (2014). Optimization of mass transfer separation elements of columnar equipment for natural gas preparation. *Chemical and Petroleum Engineering*, 49(11-12), 736-740.

[7] Voynov, N., Nikolaev, N., Kustod, F., Nikolaev, A., Tarovatyu, D. (2008). Vortex contact stages for rectification. *Chemistry of plant raw materials*, 3, 173-184.

[8] Akulich, A. V. (1999). Hydrodynamics of two-chamber vortex dryers with counter-colliding flows of a gaseous suspension. *Journal of Engineering Physics and Thermophysics*, 72(3), 392-395.

[9] Kaewklum, R., Kuprianov, V. I., Douglas, P. L. (2009). Hydrodynamics of air–sand flow in a conical swirling fluidized bed:

A comparative study between tangential and axial air entries. *Energy Conversion and Management*, 50, 2999-3006.

[10] Zardi, F. (2011). Vortex granulation: The new route to high-quality solid nitrogen fertilizers. *Nitrogen & Syngas conference & exhibition.*

[11] Pat. № 2013/0316078 A1, US, Int. Cl C07C 273/02. *Fluid bed granulation of urea and related apparatus* / Gianfranco Benedetti; assignee Urea Casale S. A. – № 13/984733; filed 09.12.2011; issued 28.11.2013.

[12] Pat. № 7955566 B2, US, Int. Cl B01J8/18. *Fluid bed granulator* / Gianfranco Benedetti; assignee Urea Casale S. A. – № 11/939915; filed 14.11.2007; issued 07.07.2011.

[13] Pat. № 2006/094620 A1, WO, Int. Cl, B01J2/16. *Process and device for fluidized bed granulation* / Gianfranco Benedetti, Carla Burlando; assignee Urea Casale S.A. – № PCT/EP2006/001502; filed 20.02.2006; issued 14.09.2006.

[14] Marrone, L. Scotto, A. (2011). New Technology for Urea Solidification: Vortex Granulator & Vibrating Skin Bucket. *The Casale 3rd Customer Symposium*, paper No. 4.3, 11 p.

[15] Saikh, M. (2013). A technical note on granulation technology: a way to optimise granules. *International Journal of Pharmaceutical Sciences Review and Research*, 4, 55-67.

[16] Solanki, H., Basuri, T., Thakkar, J., Patel, C. (2010). Recent advances in granulation technology. *International Journal of Pharmaceutical Sciences Review and Research*, 48-54.

[17] Patel, P., Telange, D., Sharma N. (2011). Comparison of Different Granulation Techniques for Lactose Monohydrate. *International Journal of Pharmaceutical Sciences and Drug Research*, 3, 222-225.

[18] Artyukhov, A. E., Fursa, A. S., Moskalenko, K. V. (2015). Classification and separation of granules in vortex granulators. *Chemical and Petroleum Engineering*, 51(5-6), 311-318.

[19] Artyukhov, A. E., Sklabinskyi, V. I. (2015). Theoretical analysis of granules movement hydrodynamics in the vortex granulators of ammonium nitrate and carbamide production. *Chemistry & chemical technology*, 9(2), 175-180.

[20] Artyukhov, A. E., Sklabinskyi, V. I. (2015). Hydrodynamics of gas flow in small-sized vortex granulators in the production of nitrogen fertilizers. *Chemistry & chemical technology*, 9(3), 337-342.
[21] Artyukhov, A. E., Sklabinskyi, V. I. (2013). Experimental and industrial implementation of porous ammonium nitrate producing process in vortex granulators. *Naukovyi Visnyk Natsionalnoho Hirnychoho Universytetu*, 6, 42-48.
[22] Artyukhov, A. (2016). Application software products for calculation trajectories of granules movement in vortex granulator. *CEUR Workshop Proceedings,* 1761, 363-373.
[23] Artyukhov, A., Sklabinskiy, V., Ivaniia, A., Moskalenko, K. (2016). Software for calculation of vortex type granulation devices. *CEUR Workshop Proceedings*, 1761, 374-385.
[24] Artyukhov, A. E., Voznyi, A. A. (2016). Thermodynamics of the vortex granulator's workspace: the impact on the structure of porous ammonium nitrate. *6th International Conference Nanomaterials: Application & Properties,* 5(2), 02NEA01.
[25] Artyukhov, A. E. Kinetics of heating and drying of porous ammonium nitrate granules in the vortex granulator *6th International Conference Nanomaterials: Application & Properties,* 5(2), 02NEA02.
[26] Artyukhov, A. E., Obodiak, V. K., Boiko, P. G., Rossi P. C. (2017). Computer modeling of hydrodynamic and heat-mass transfer processes in the vortex type granulation devices. *CEUR Workshop Proceedings*, 1844, 33-47.
[27] Artyukhov, A. E., Sklabinskyi, V. I. (2017). Investigation of the temperature field of coolant in the installations for obtaining 3D nanostructured porous surface layer on the granules of ammonium nitrate. *Journal of Nano- and Electronic Physics*, 9(1), 01015-1-01015-4.
[28] Levchenko, D. O., Artyukhov, A. E., Yurko, I. V. (2017). Maisotsenko cycle applications in multi-stage ejector recycling module for chemical production. *IOP Conference Series: Materials Science and Engineering,* 233(1), 2017, article number 012024.

[29] Artyukhov, A. E. (2017). Internal Recycled Particles Circulation in Vortex Granulator. *Chemical and Petroleum Engineering*, 53(7-8), 423-429.

[30] Artyukhov, A. E., Ivaniia, A. V. (2017). Obtaining of porous ammonium nitrate in multistage and multifunctional vortex granulators. *Naukovyi Visnyk Natsionalnoho Hirnychoho Universytetu*, 6, 68-75.

[31] *Aspen HYSYS®* [E-text type]. Location of document: http://www.aspentech.com/products/aspen-hysys/.

[32] *SimSci PRO/II* [E-text type]. Location of document: http://software.schneider-electric.com/products/simsci/design/pro-ii/.

[33] *GasCondOil* [E-text type]. Location of document: http://thermogas.kiev.ua/Products.htm.

[34] *ChemCAD* [E-text type]. Location of document: http://www.chemstations.com.

[35] *DESIGN II* [E-text type]. Location of document: https://www.winsim.com/design.html.

[36] *SIGMAFINE* [E-text type]. Location of document: https://sigmafine.pimsoftinc.com.

[37] *gPROMS ModelBuilder* [E-text type]. Location of document: https://www.psenterprise.com/products/gproms.

[38] *GIBBS* [E-text type]. Location of document: http://www.gibbsim.ru.

[39] *ANSYS Fluent* [E-text type]. Location of document: http://www.ansys.com/Products/Fluids/ANSYS-Fluent.

[40] *COSMOS FlowWorks* [E-text type]. Location of document: http://www.solidworks.com/sw/products/simulation/flow-simulation.htm.

[41] *FlowVision* [E-text type]. Location of document: http://tesis.com.ru.

[42] *Flow-3D CFD* [E-text type]. Location of document: https://www.flow3d.com.

[43] *Ansys CFX* [E-text type]. Location of document: http://www.ansys.com/products/fluids/ansys-cfx.

[44] *COMSOL Multiphysics* [E-text type]. Location of document: https://www.comsol.com.

[45] Crowe, C. T. (2006). *Multiphase flow handbook*. New York: Taylor & Francis Group, LLC.

[46] Zimmermann, S., Taghipour, F., (2005). CFD Modeling of the Hydrodynamics and Reaction Kinetics of FCC Fluidized-Bed Reactors. *Industrial an Engineering Chemistry Research*, 44 (26), 9818-9827.

[47] Li, X., Li, G., (2016). Modeling and Simulation of Hydrodynamic Bubble-Liquid Turbulent Flows in Bubble-Column Reactors. *Chemical Engineering & Technology*, 39(12), 2380-2388.

[48] Wang, T., (2011). Simulation of bubble column reactors using CFD coupled with a population balance model. *Frontiers of Chemical Science and Engineering*, 5(2), 162-172.

[49] Tahvildari, K., Tavakoli, H., Mashayekhi, A, Golmohammadzadeh, R. (2016). Modeling and simulation of membrane separation process using computational fluid dynamics. *Arabian Journal of Chemistry*, 9(1), 72-78.

[50] Nouh, S. A., Lau, K. K. Shariff, A. M. (2010). Modeling and Simulation of Fixed Bed Adsorption Column using Integrated CFD Approach. *Journal of Applied Sciences*, 10, 3229-3235.

[51] Zhang, L., Li, Z., Yang, N., Jiang, B., Pavlenko, A. N., Volodin, O. A. (2016). Hydrodynamics and mass-transfer characteristics analysis of vapor-liquid flow of dual-flow tray. *Journal of Engineering Thermophysics*, 25(4), 449-463.

[52] Chen, Q., Zhai, Z., Wang, L. (2007). Computer modeling of multiscale fluid flow and heat and mass transfer in engineered spaces. *Chemical Engineering Science*, 62(13), 3580-3588.

[53] Wang, Z., Xing, Y., Liu, X., Zhao, L., Ji, Y. (2016). Computer modeling of droplets impact on heat transfer during spray cooling under vibration environment. *Applied Thermal Engineering*, 107 (25), 453-462.

[54] Trojan, M. (2015). Computer modeling of a convective steam superheater. *Archives of thermodynamics*, 36(1), 125-137.

[55] Baukal, C. E. Jr., Gershtein V., Xianming, J. L. (2000). *Computational Fluid Dynamics in Industrial Combustion*. Boca Raton : CRC Press.

[56] Artyukhov, A. E., Sklabinskyi, V. I. (2016). 3D nanostructured porous layer of ammonium nitrate: influence of the moisturizing method on the layer's structure. *Journal of Nano- and Electronic Physics*, 8(3).04051-1-04051-5.

[57] Artyukhov, A. E., Sklabinskyi, V. I. (2016). Thermodynamic conditions for obtaining 3D nanostructured porous surface layer on the granules of ammonium nitrate. *Journal of Nano- and Electronic Physics*, 8(4), 04083-1-04083-5.

[58] Artyukhov, A., Voznyi, A. (2017). The meso- and macropores generating process via modification of NH_4NO_3 granules: thermodynamic regime impact on the granule structure. *6th International Conference Nanomaterials: Application & Properties*, 6(3), 03NEA09.

[59] Artyukhov, A., Ivaniia, A. (2017). Experimental and industrial introduction of the improved devices to generate NH_4NO_3 granules with nanoporous structure. *6th International Conference Nanomaterials: Application & Properties*, 6(3), 03NEA10.

[60] Artyukhov, A., Artyukhova, N., Ivaniia, A., Galenin, R. (2017). Progressive equipment for generation of the porous ammonium nitrate with 3D nanostructure. *6th International Conference Nanomaterials: Application & Properties*, 6(3), 03NEA06.

[61] Artyukhov, A., Artyukhova, N., Ivaniia, A., Gabrusenoks, J. (2017). Multilayer modified NH_4NO_3 granules with 3D nanoporous structure: effect of the heat treatment regime on the structure of macro- and mezopores. *IEEE International Young Scientists Forum on Applied Physics and Engineering*, 315-318.

In: Advances in Engineering Research
Editor: Victoria M. Petrova
ISBN: 978-1-53614-218-1

Chapter 3

Piezoelectric Controlled Ceramics with Metallic Implants: A New Alternative for the Engineering Field

Ernesto Suaste-Gómez**, Daniel Hernández-Rivera and Grissel Rodríguez-Roldán
Department of Electrical Engineering, Section of Bioelectronics, Center for Research and Advanced Studies of the National Polytechnic Institute, Mexico City, Mexico

Abstract

Piezoelectric materials were discovered many years ago; however, new compounds and fabrication techniques are continually explored in order to get the optimal and new properties. In recent years, a new piezoelectric ceramic with Platinum implant has been fabricated. The application possibilities of this ceramic are increased since there is a third electrode available which can be used to control its behavior. This device

---

* Corresponding Author address: Email: esuaste@cinvestav.mx.

is called Piezoelectric Controlled Ceramic (PCC). The PCCs were made of PLZT and $BaTiO_3$ by the oxide-mixing technique, during this process; a platinum wire of 300 μm diameter was implanted in the middle of the ceramic bulk in a transversal way. One important characteristic of the PCC is that its dielectric constant can be modified by introducing an alternate current signal in the implant. On the other hand, the response of the ceramic to light or mechanical stimuli has superior sensibility in the area near the implant. The PCC has been used in a wide variety of applications in the engineering field, such as opacity and mechanical vibration sensors. It was successfully used as a heart beat sensor (piezoplethysmography). Additionally, it has been employed as frequency modulator for wireless transmission and analog filter. The PCC with two implants has shown good results as insulation device in electrocardiography. Although the versatility of the PCC in the engineering field has been demonstrated, it is necessary to study more deeply the phenomena involved in its behavior.

Keywords: piezoelectricity, ferroelectric ceramic, transducer, sensor

INTRODUCTION

At present, there is a substantial demand of products based on piezoelectric materials for use in devices such as: sensors, accelerometers, gyroscopes, filters, resonators, transformers, micro-motors, electronic ignition elements, among others. These materials offer a wide variety of applications due to their spontaneous polarization, since they do not require a power supply for their use. It is also possible to take advantage of their high electrical impedance or the dependence of their dielectric behaviour. Then, the benefits of piezoelectric materials are not only limited to their use as transducers, but also, as isolators.

In order to find new application opportunities, it is necessary to develop new manufacturing techniques, new materials, as well as combinations of these that allow obtaining new properties or improving existing ones. In this context, a new device called piezoelectric controlled ceramic has been developed. This design is based on the insertion of a Platinum wire into the body of a conventional piezoelectric ceramic [1].The Pt implant concentrates a greater amount of domains than the rest of the body of the ceramic when it is polarized. This characteristic allows a free flow of electric charges when one of the faces of the ceramic is stimulated electrically, optically, thermally or mechanically [1-4]. The implant can work as an third electrode. This characteristic allows to free one of the faces of the ceramic to be used as a sensing area. On the other hand, the implant in the PCC can also work as an input element. Electric charges flow across the implant, modifying the dielectric constant of the ceramic. This fact opens a wide variety of applications because many properties of the piezoelectric ceramics depend on their permittivity.

This chapter presents a compendium of the characterization and examples of applications that have been developed based on the PCC. It covers mechanical, optical, thermal, biomedical, electrical and electronic aspects, showing evidence of its high potential of application.

PIEZOELECTRIC CONTROLLED CERAMIC

Piezoelectricity

The piezoelectric effect was discovered by the Curies in 1880. This effect is characterized by the appearance of electric charges on the faces of a crystalline material when it is subjected to a mechanical pressure. Its counterpart, consists of the appearance of mechanical deformations when it is applied an electric field [5-7]. A general outline of this phenomenon is shown in Figure 1.

Only the materials that belong to 20 of the 21 crystalline, that has no center of symmetry present piezoelectric properties. For a long time natural crystals such as quartz, tourmaline or Rochelle salt dominated the field of piezoelectricity [5-7]. However, over time and in response to the need for better materials, piezoelectric ceramics were developed. Piezoelectric ceramics are manufactured in different ways depending on the application to which they are destined. Piezoelectric ceramics are classified mainly by their chemical composition. The most common classification criteria include performance parameters, as well as their behavior under electrical, optical, thermal and mechanical stimuli [5-7]. Among the main piezoelectric materials used today are $BaTiO_3$, $(Bi_{0.5}Na_{0.5})_{0.935}Ba_{0.065}TiO_3$ (BNBT), $KNbO_3$, $PbZrO_3$, PLZT, PZT and $PbTiO_3$.

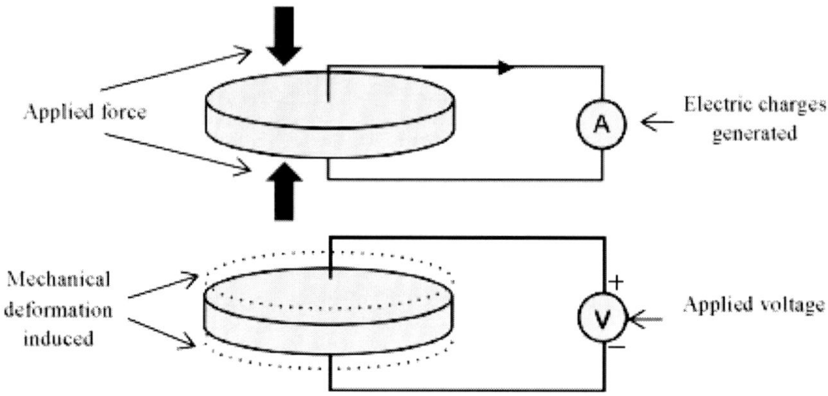

Figure 1. Piezoelectric effect.

Piezoelectric Controlled Ceramic

In this new device called PCC, a platinum wire was implanted in a conventional piezoelectric ceramic during the manufacturing process. The aim of this fact is to obtain a third electrode that allows the entry and exit of electrical charges in the ceramic [1]. The PCCs of $BaTiO_3$ and PLZT were manufactured. Platinum was chosen as implant material because its high resistance to chemical attack, excellent characteristics at high temperature (melting point of 1,768.3 °C) and its stable electrical properties. Figure 2 shows a PCC scheme with platinum implant.

Alternatively, a PCC with two implants as shown in Figure 3 was fabricated. Also, the three different symbols illustrated in Figure 4 were proposed in order to describe schematically the different applications.

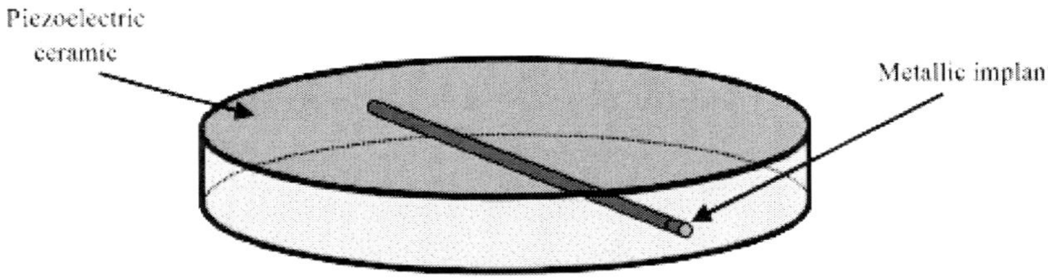

Figure 2. Piezoelectric controlled ceramic with one implant.

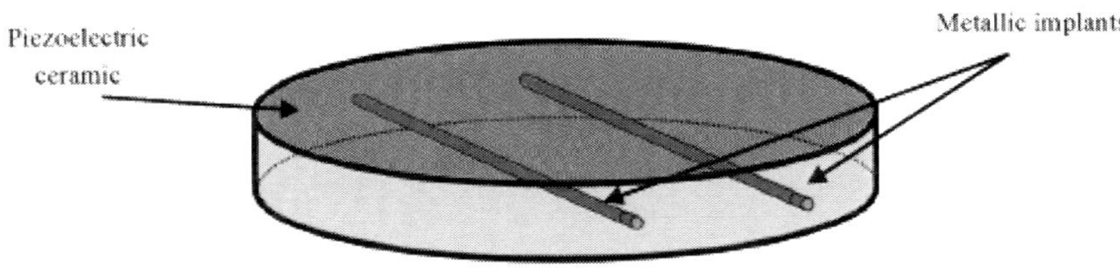

Figure 3. Piezoelectric controlled ceramic with two implants.

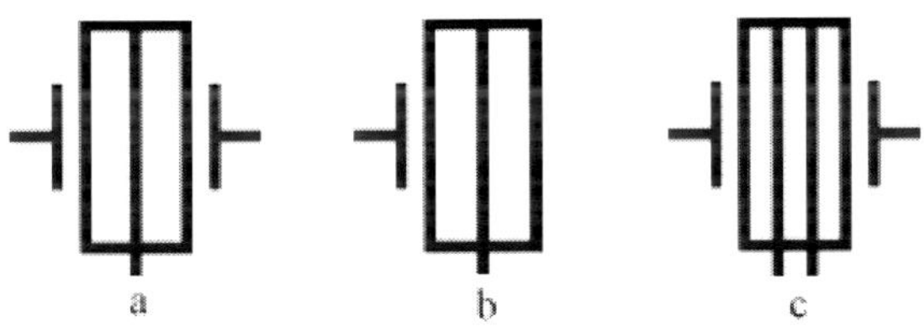

Figure 4. Schematic diagrams of PCC. a) PCC with one implant and two electrodes, b) with one implant and one electrode and c) with two implants and two electrodes.

Operating Principle

When the PCC is polarized, the mobile charges in the Pt wire accumulate on the surface until the produced field cancels the externally applied field. This produces equilibrium as it is shown in Figure 5 [1].

The wire of Pt has a singular performance in the PCC because the domains of the ceramic cover it. This type of domain wall separates the polarized domains perpendicularly. When the PCC is polarized, the electric field exerts an opposite torque on the polar axis of opposite domains until the polarization rotates permanently.

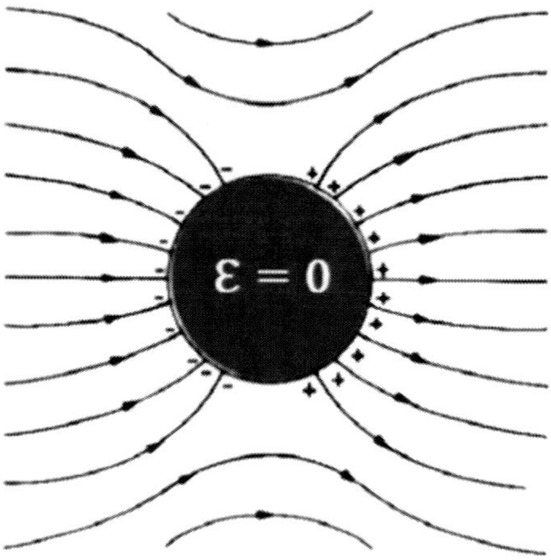

Figure 5. Electric field waves across of Pt-wire.

High concentrations of domains are produced on the platinum wire, as it is seen in Figure 6. These concentrations achieve a free flow of charges around the Pt when one of the faces of the ceramic is stimulated electrically, optically, thermally or mechanically. This results in a greater sensitivity in the area near the implant [1, 2, 4].

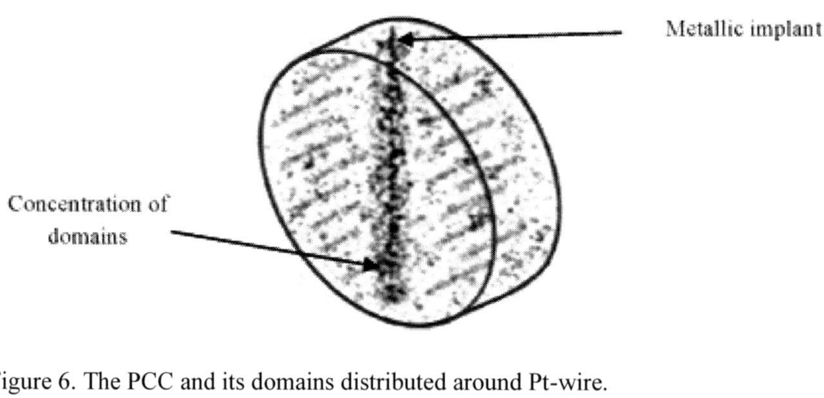

Figure 6. The PCC and its domains distributed around Pt-wire.

CHARACTERIZATION

Fabrication

The oxide-mixing route is the most common method to fabricate piezoelectric ceramics. This method involves mixing and grinding, calcinations, grinding, shaping, densification and finishing [7].

The PLZT ((9/65/35) and $BaTiO_3$ceramics with implant were prepared by the conventional ceramic processing via oxide mixing route. An example of the fabricated ceramics can be seen in Figure 7.

Firstly, the raw materials were mixed in an automatic ball mill (Pulverisette 2, Fritsch) for 20 minutes using polyvinyl alcohol with a rate of 1.5 drops per gram of mixture.

Then, powders were pressed at 3,500 kg/cm^2. During this process, a 0.3 mm diameter Pt-wire was inserted in the middle of the ceramic in a longitudinal way.

Discs of 10 mm diameter and 2 mm of thickness with a metallic electrode totally immersed in the ceramic were obtained. Alternatively, a ceramic with two implants was fabricated.

The ceramics were sintered at 1200 °C in an air-atmosphere furnace using a platinum crucible. After the sintering process, silver paint electrodes were deposited on both faces of the PCC.

Finally, the samples were electrically poled at 1.5 kV/mm for one hour at 80 °C in a silicone oil bath.

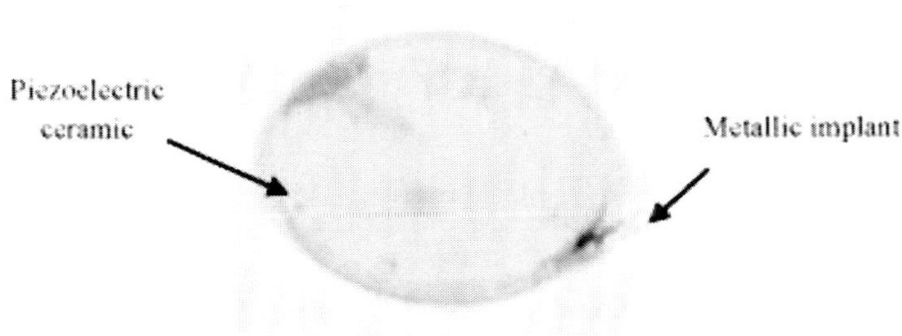

Figure 7. PCC with implant.

Microscopy and Microanalysis

In order to obtain topographical, morphological and compositional information, samples were evaluated by scanning electron microscopy (SEM). The micrographs of the longitudinal part of the PCC of $BaTiO_3$ were obtained using a SEM-EDS JEOL model: JSM-7401F at a scale of 100 μm. An EDS analysis was performed to check if there was any reaction that could alter the chemical composition of the ceramic at the border with the implant. The points at which the analysis was made as well as the results thereof are presented in Figures 8 and 9 respectively.

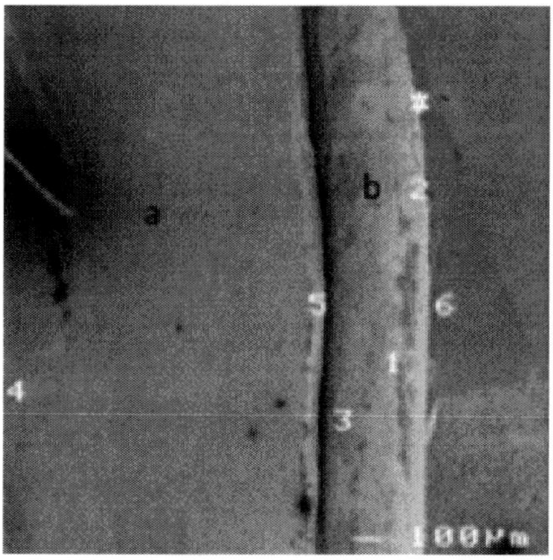

Figure 8. Micrograph of PCC. a) ceramic ($BaTiO_3$) and b) implant (Pt).

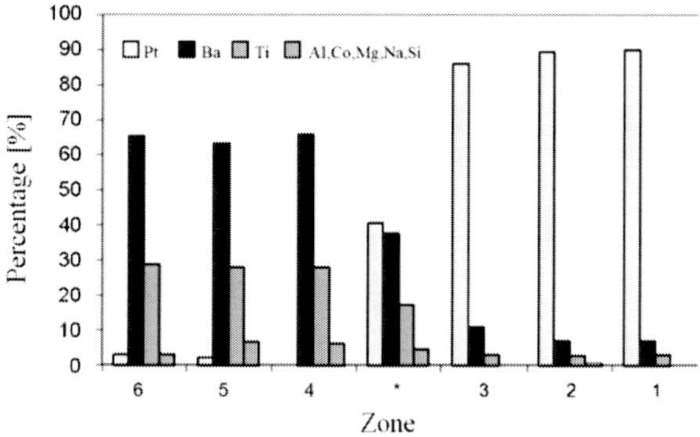

Figure 9. EDS analysis of $BaTiO_3$ with implant.

From the SEM micrograph it can be seen that there is a clear boundary between the Pt and the $BaTiO_3$. It is possible to appreciate a uniform and compact structure in the area surrounding the implant. Regarding the EDS analysis, a uniform distribution of Pt and $BaTiO_3$ was observed. However, traces of other elements were also found. This fact may be due to the impurities contained in the raw material.

Curie Temperature

Curie temperature (T_C) is a parameter of great importance in ferroelectric materials. For this reason, it was necessary to know how the platinum implant alters the T_C in the composite used.

For this purpose, the dielectric constant ε_r was obtained as a function of temperature for $BaTiO_3$ and PLZT ceramics with and without platinum implants. The T_C was determined by means of the inflection point of the permittivity curve.

Figure 10 shows the dependence of ε on the temperature for the ceramics of $BaTiO_3$ with and without platinum implant. The T_C of the ceramic without implant is 130 °C, which corresponds to the literature [7]. Regarding the ceramic with implant, the dielectric constant decreased by 28% with respect to the ceramic without implant. The T_C did not have significant differences between ceramics with and without implants.

The behavior of the dielectric constant with respect to temperature was also obtained for PLZT ceramics with and without implants. Figure 11 shows that the T_C was 360 °C for the ceramic with implant and 375 °C for the ceramic without implant, both within the acceptable range for PLZT ceramics. On the other hand the dielectric constant was lower in the ceramic with implant.

The electrical charges generated by the mechanical stimuli were picked up by a current amplifier and recorded by an oscilloscope. The corresponding graph is shown in Figure 13. The magnitude of points B, E, H, (points located directly on the implant) is 42% higher than the rest of the points evaluated.

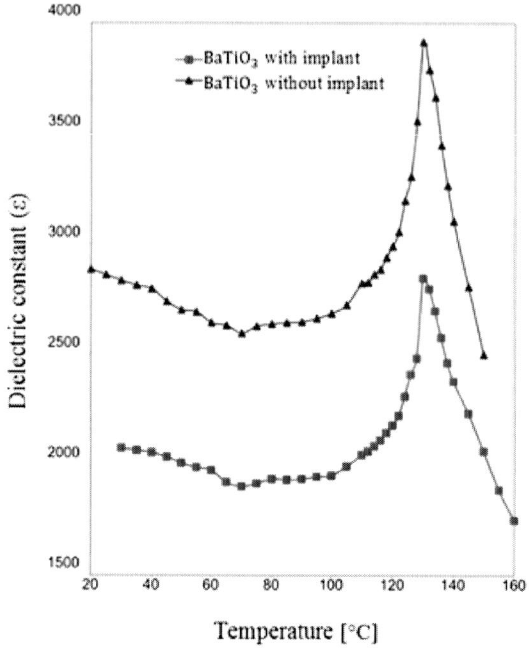

Figure 10. Dielectric constant of $BaTiO_3$ ceramics with and without Pt-wire.

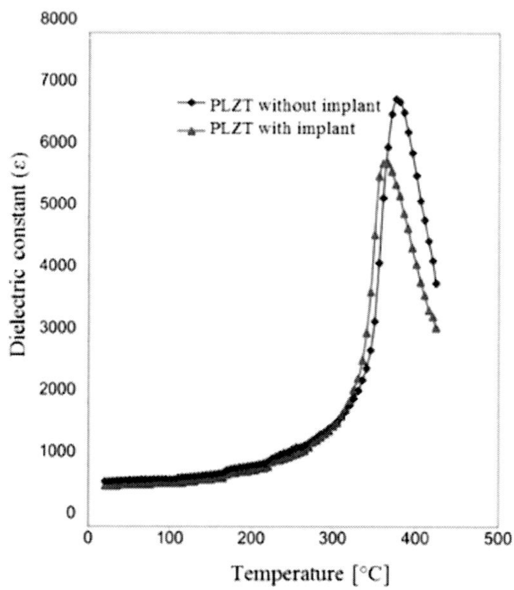

Figure 11. Dielectric constant of PLZT ceramics with and without Pt-wire.

Electrical Response to Mechanical Stimuli

The direct piezoelectric effect was obtained by applying mechanical stimuli in 9 different areas of the controlled ceramic as shown in Figure 12. The force applied for this test was 53.9 mN at a frequency of 2.2 Hz.

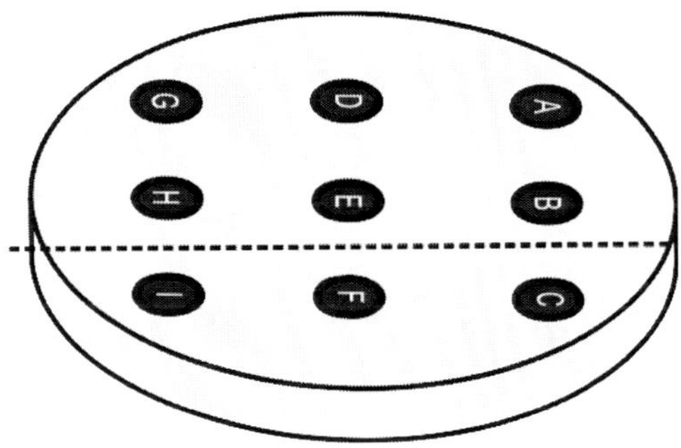

Figure 12. Zones selected of PCC for optical and mechanical analysis.

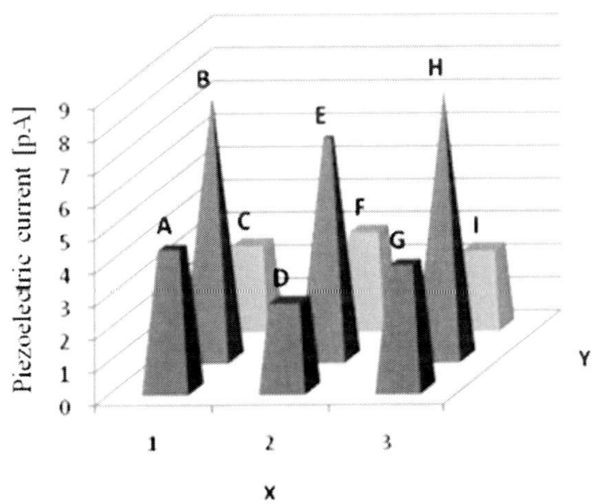

Figure 13. Electrical response of the PCC related to mechanical stimuli.

Electrical Response to Light Stimuli

Piezoelectric ceramics also respond to light stimuli. For this characterization, a PCC with one of the free faces was used to make a laser light impinge. Figure 14 shows the amplitude of the photovoltaic signal, due to the excitation of the PLZT-type ceramic in the same points shown in Figure 12. The ceramic was stimulated by a beam of light (650 nm) modulated at 3 Hz. The points located directly on the implant have a greater amplitude than those points away from it. This is because there is a higher concentration of ferroelectric domains around the implant.

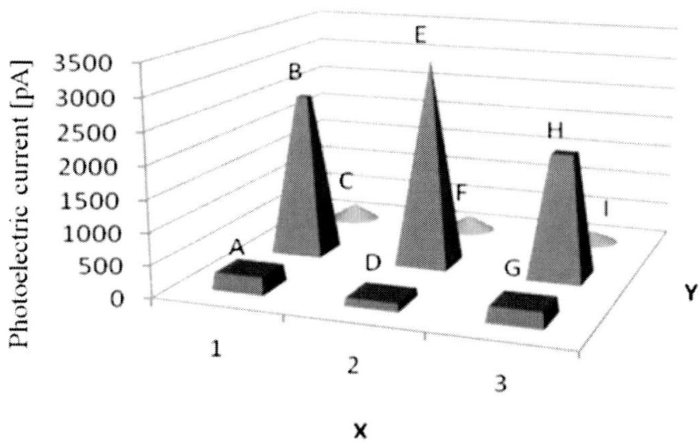

Figure 14. Electrical response of the PCC related to light stimuli.

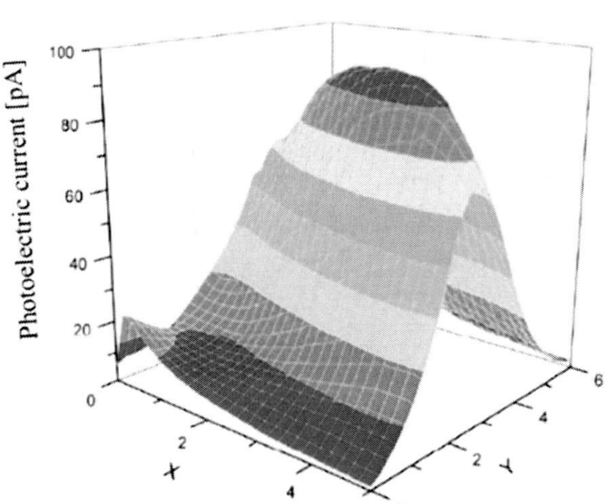

Figure 15. High resolution response of the PCC related to light stimuli.

Additionally, the graph shown in Figure 15 was obtained with a high resolution. Similarly, the points on the implant have a greater amplitude than those points away from it.

Electrical Insulation of the PCC with Two Implants

An alternative to piezoelectric controlled ceramics consists of using two implants. In this case, the two faces of the ceramic and the two implants can work as electrodes separated by a high electrical impedance.

Three experimental setups (high voltage, DC resistance, leakage current) were used in order to determine the electrical isolation characteristics of the PCC with two implants. In the first experimental setup an isolation test was made applying a DC electric field through Pt-wires and one lateral face. Firstly, the PCC2 was immersed into silicone oil bath with the purpose of eliminate any sparks and then the electric field was increased by 500 V each minute. A high voltage probe was used for monitoring when the ceramic fails (short circuit), as shown in Figure 16.

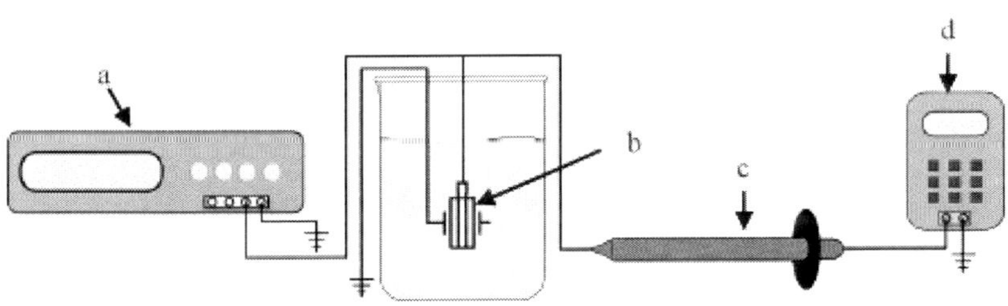

Figure 16. High voltage measured in an experimental setup to determine isolation level of PCC2. a) High voltage supply, b) PCC2 immersed into silicone oil bath, c) high voltage probe, d) voltmeter.

In order to determine electrical resistance values from different parts of the PCC2 a second experimental test was made. This test was achieved with a collaboration of the National Metrology Center (CENAM, México). Figure 17 shows the points of interest for this test.

This procedure was made at room temperature using an electrometer Mod. 6517B (Keithley). To measure PCC2 resistance between A, B, C, and D parts, a PTFE (Teflon) chamber was used. The chamber showed in Figure 18 contains high purity coaxial cables of 40 cm long with Teflon isolation, a thermocouple to measure the internal temperature and a metallic cage to isolate electromagnetically the chamber (as Faraday cage).

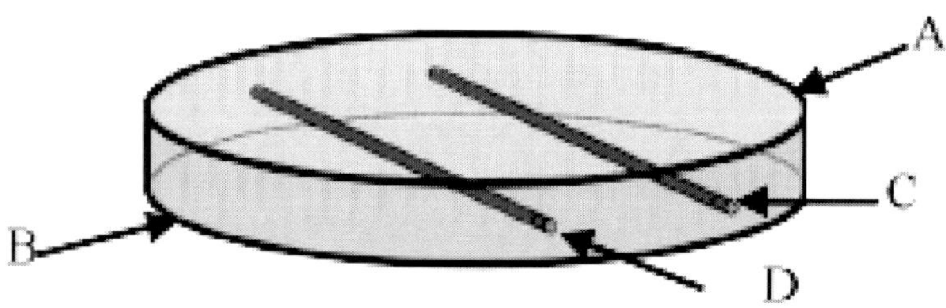

Figure 17. Scheme of PCC and four resistance parts. A, B represents lateral sides of PCC (top and bottom respectively) and C, D represents internal Pt-wires.

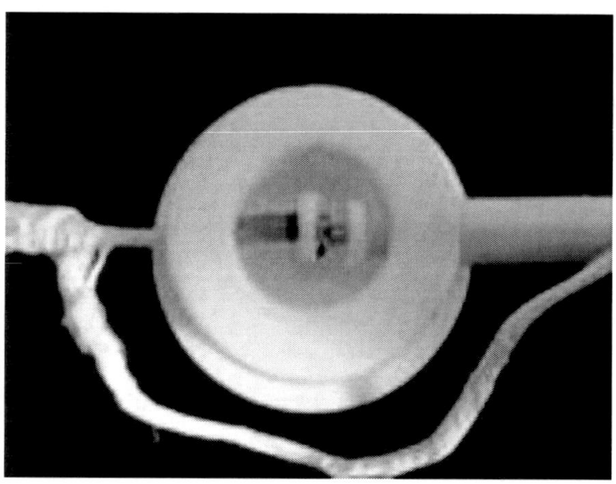

Figure 18. The Teflon chamber is connected to the PCC with its coaxial cables.

The third experimental setup shown in Figure 19 was used to measure the PCC leakage. A voltage signal applied to the PCC2 and the resulting leakage current through the ceramic were measured simultaneously. The resistor limits the current and it also helps to reduce noise. A typical value for the resistor is 1 MΩ, although this resistance can be decreased for larger capacitance values of PCC2.

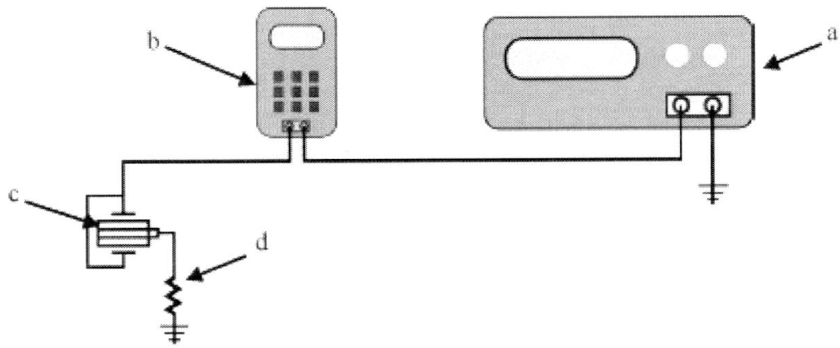

Figure 19. Experimental setup to measure leakage current in PCC. a) Voltage power supply, b) ampere meter, c) PCC2, d) resistor.

In the first test, the PCC2 fails at 6 kV. With respect to the second experimental test, the resistance of the PCC2 device between A and B parts was 3.9×10^9 Ω and between lateral and internal electrodes the resistance was 1.8×10^9 Ω. Finally, the leakage current in the third test was 6 nA.

APPLICATIONS

Analogue Electronics

Current Source

A current source based on a PCC implanted with two Pt wires was proposed. This configuration allowed to obtain a controlled electric current (12 μA at 90 μA) with a conventional function generator. For this, a 2.086 MHz sinusoidal signal with a variable peak-to-peak amplitude of 680 mV to 6 V was used. The experimental arrangement is shown in Figure 20 and the response can be seen in Figure 21.

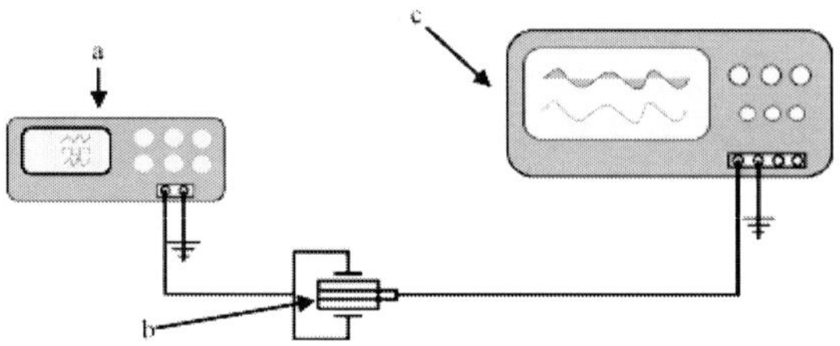

Figure 20. Experimental setup of current source. a) Function generator, b) PCC and c) oscilloscope.

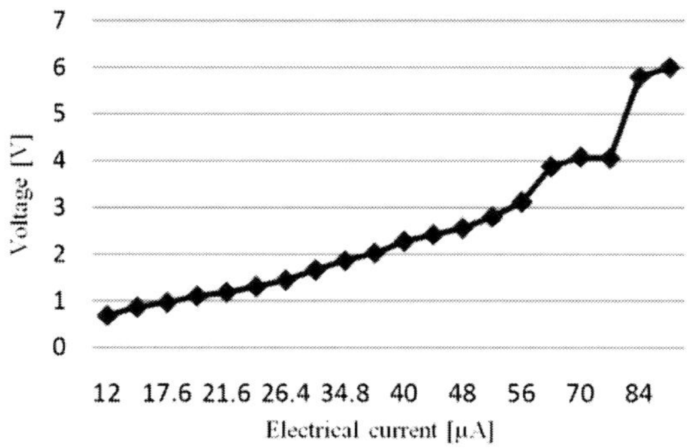

Figure 21. Response of current source based on PCC2.

High Frequency Oscillator

The PCC can also work as a high frequency oscillator in a parallel resonator circuit as shown in Figure 22. With this configuration, a signal of 41.66 MHz was obtained at the output of the digital inverter. The resulting graph can be seen in Figure 23.

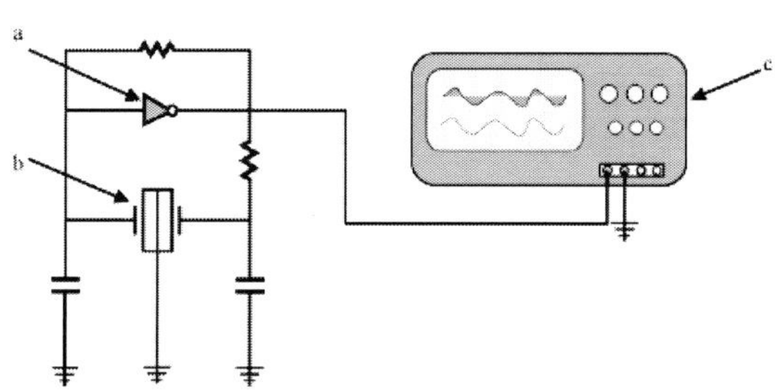

Figure 22. PCC used as oscillator circuit, a) TTL inverter, b) PCC and c) oscilloscope.

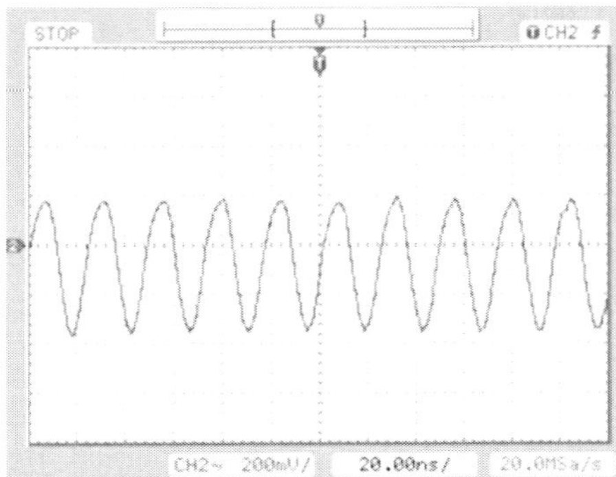

Figure 23. Signal obtained of the oscillator circuit.

Signal Superpositioner

In this case, the PCC works as an oscillator in a parallel resonator circuit where the platinum implant works as an input element. A superposition of the signal generated by the resonator circuit (Figure 24) and low frequency signals (sine waves, triangular waves and square waves) introduced into the implant was obtained. The resulting signals are observed in Figure 25.

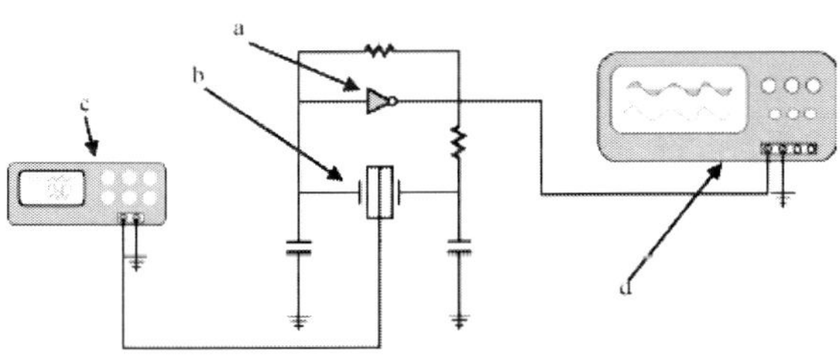

Figure 24. PCC used as oscillator circuit with a control signal. a) TTL inverter, b) PCC, c) function generator and d) oscilloscope.

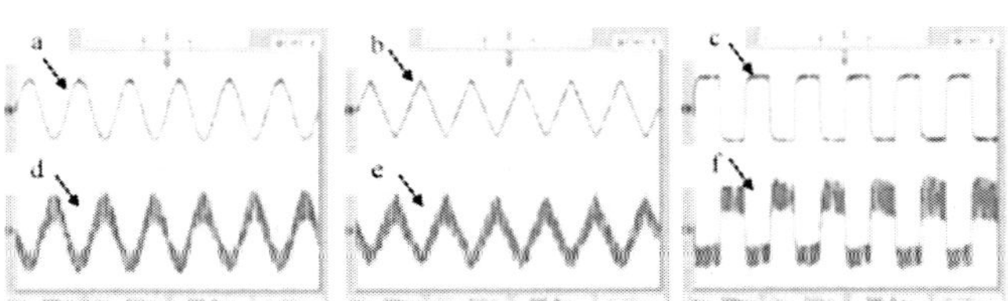

Figure 25. Graphs of PCC used as overlapping circuit. a) sine wave, b) triangular wave and c) square wave at 10 kHz and (d, e, f) overlapping signals obtained of the inverter circuit.

Amplitude Modulator

In the experimental arrangement shown in Figure 26, two oscillating signals obtained from a signal generator were introduced on the faces of the PCC. The signal resulting from the combination of both was acquired by means of the implant. Figure 27 shows the signal modulated in amplitude obtained with this configuration.

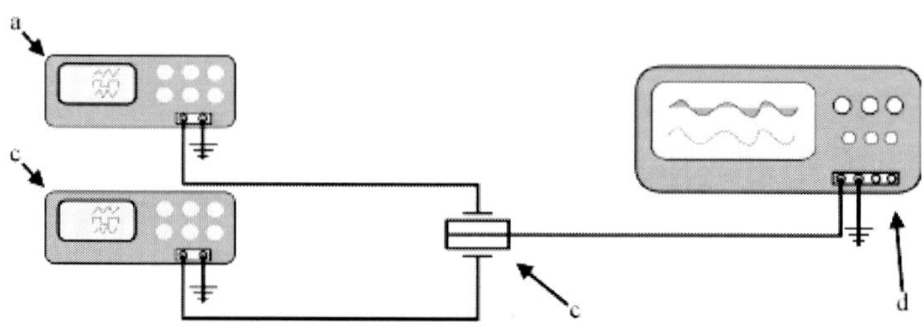

Figure 26. PCC as signal modulator. (a,b) Function generator, d) PCC and e) oscilloscope.

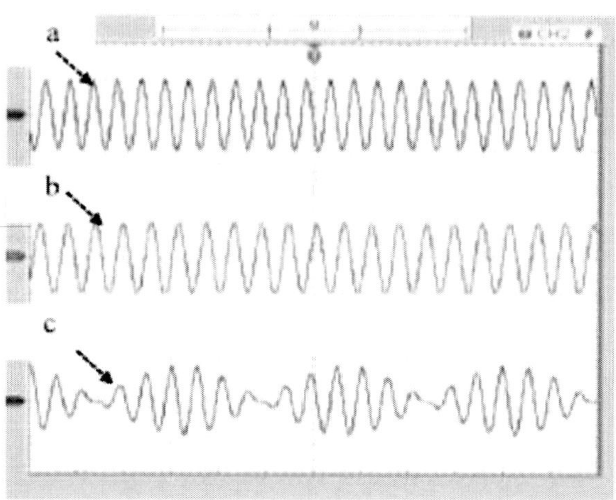

Figure 27. Response of PCC used as amplitude modulator. a) sinewave input at 1 kHz, b) sinewave input at 2 kHz and c) output signal.

Frequency Modulator

The dielectric constant of the PCC can be modified by supplying electrical charges in the Pt implant. Considering the piezoelectric ceramic as a circular capacitor with parallel plates, the dielectric constant can be obtained according equation (1):

$$\varepsilon = \frac{Cd}{\varepsilon_0 A} = \frac{Cd}{\varepsilon_0(\pi * r^2)} \tag{1}$$

Where, ε_0 is the vacuum dielectric constant (8.85×10^{-12} F / m), d the distance between the faces of the ceramic and the A area of the faces. Therefore, the dielectric constant of the ceramic is directly proportional to the capacitance. In this way we can know that certain electric charges supplied in the Pt implant cause variability in the capacitance presented by the ceramic.

Taking advantage of the aforementioned behavior, any oscillator circuit where the frequency depends directly on a capacitance, can be used in conjunction with the PCC as a frequency modulator. The piezoelectric ceramic (in this case) works as a variable capacitor controlled by the signal introduced to the implant.

In this case, the capacitor that controls the oscillation frequency of a 555 timer IC was replaced by the PCC as shown in Figure 28. By exciting the PCC with a signal in the metallic implant, the capacitance of the ceramic is modified. This fact causes the output signal of the astable oscillator has a variable frequency as shown in Figure 29.

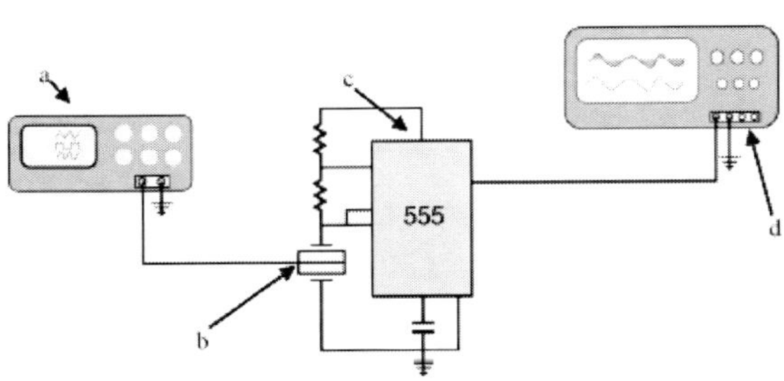

Figure 28. PCC used as frequency modulator. a) Function generator, b) PCC, c) I.C. 555 and e) oscilloscope.

Later, an electrocardiographic signal was applied to the implant of the PCC, obtaining as a result, a biological signal modulator (Figure 30). The ECG signal induced a change in the PCC permittivity; as a consequence, the oscillator frequency was modified proportionally to the ECG signal. Using a frequency to voltage converter was possible to recover the original signal applied to the PCC. Figure 31 shows the ECG signal applied to the PCC and the ECG signal recovered by the PCC modulator.

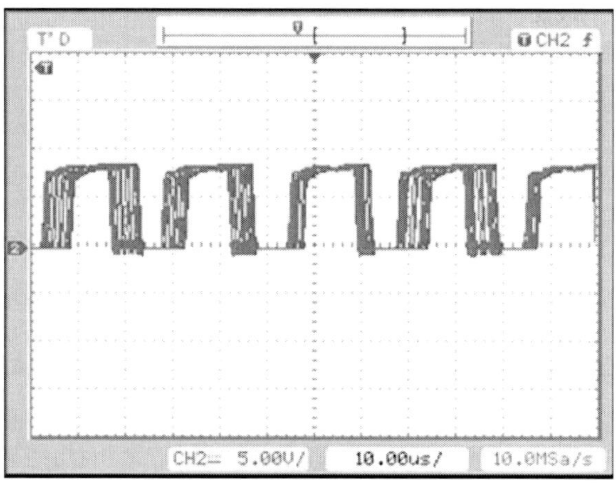

Figure 29. Square wave with variable frequency.

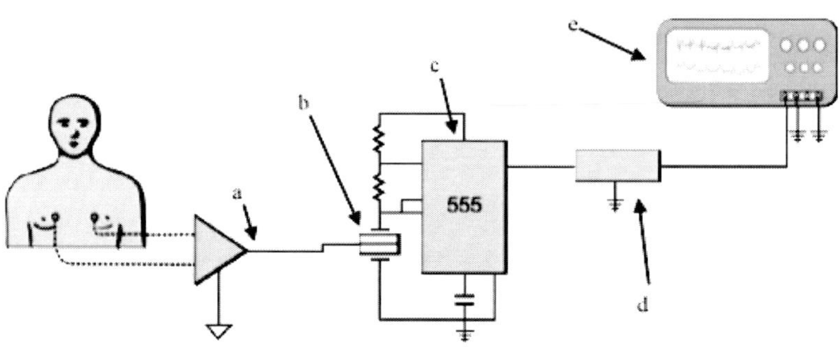

Figure 30. PCC as modulator for ECG. a) ECG signal, b) PCC, c) I.C. 555 and d) frequency to voltage converter and e) oscilloscope.

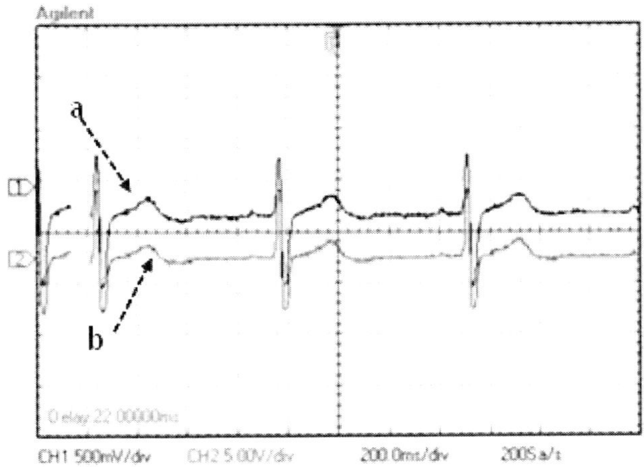

Figure 31. a) Input ECG signal and b) ECG recovered signal after demodulation.

Isolation Device for ECG

The PCC2 is a piezoelectric controlled ceramic with 2 implants as illustrated in Figure 32. This device can be used as an isolation system since there is a very high electrical impedance between the faces of the ceramic and the metal implants.

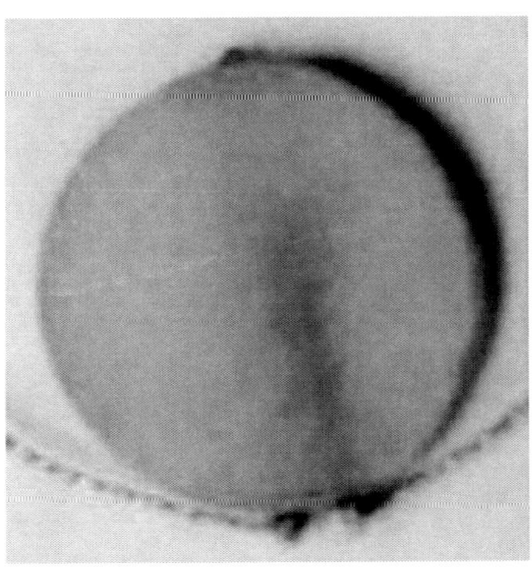

Figure 32. PCC with two implants.

For this application, PCC2 was used as an isolation system for electrocardiography. The configuration used is shown in Figure 33. It consists of an amplifier device for ECG based on an INA128 IC, a sampler circuit that converts the ECG signal into a pulsating high frequency signal (a high frequency signal is necessary for the proper functioning of the PCC), the PCC2 as an isolation element and an envelope detector to recover the signal. The input ECG signal and the signal obtained after isolation are shown in Figure 34.

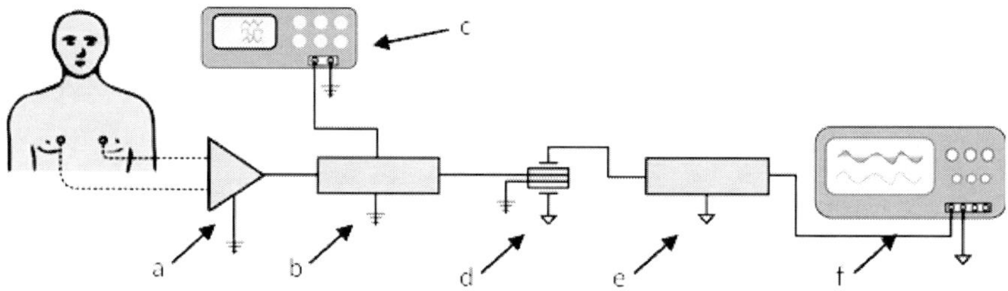

Figure 33. PCC2 as isolator in ECG. a) ECG signal, b) sampler circuit, c) function generator, d) PCC2, e) envelope detector and conditioning and f) oscilloscope.

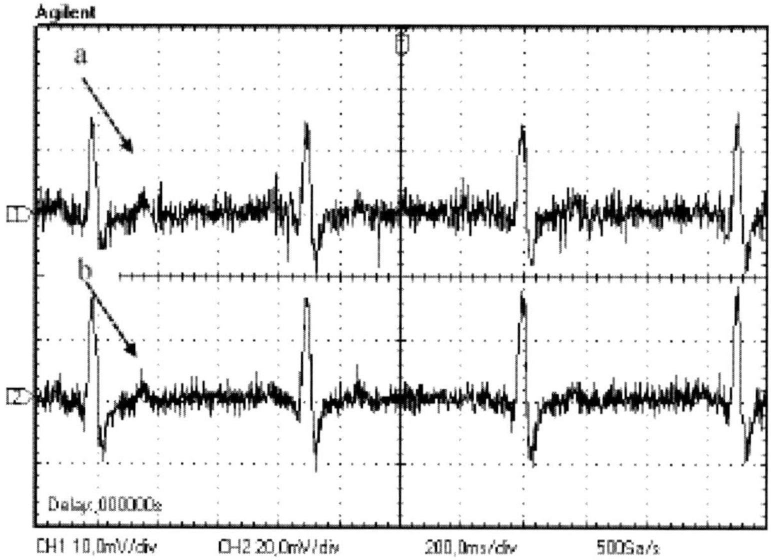

Figure 34. a) ECG signal and b) ECG signal recovered after of the isolator.

Analog Filters

A piezoelectric ceramic can be represented as a series RLC circuit in parallel to a capacitor C_0. Due to the type of impedance that represents a ceramic, it is possible to apply low pass or high pass filters, combining a piezoelectric ceramic and resistive elements. Considering that at frequencies below the resonance frequency, the RLC impedance is very high and has no significant effect on the value of C_0, the filter can be designed taking into account only the capacitance in parallel.

Low Pass Filter

In this configuration, the PCC implant was connected to ground and one face of the ceramic was connected to a resistive element Rf as illustrated in Figure 35. Figure 36 shows the filter response for resistors of 1 Ω, 1 MΩ and 10 MΩ. When the Rf resistance was 1 MΩ, the response obtained was that of a low pass filter with a cutoff frequency of 49.13 Hz.

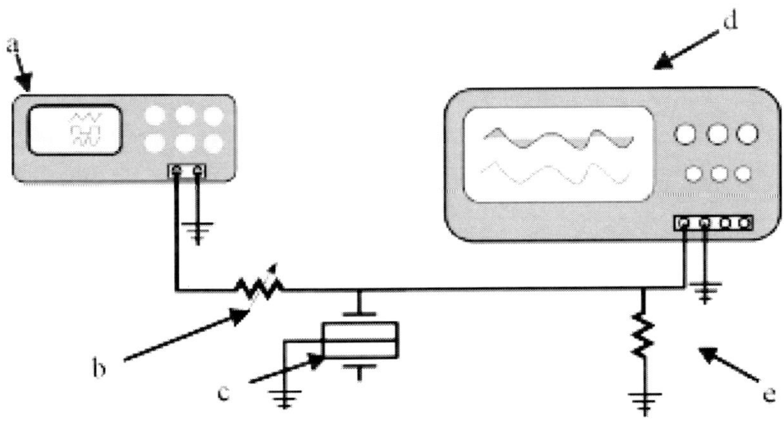

Figure 35. PCC as low pass filter. a) Function generator, b) Resistance Rf, c) PCC, d) oscilloscope and e) input impedance of the oscilloscope.

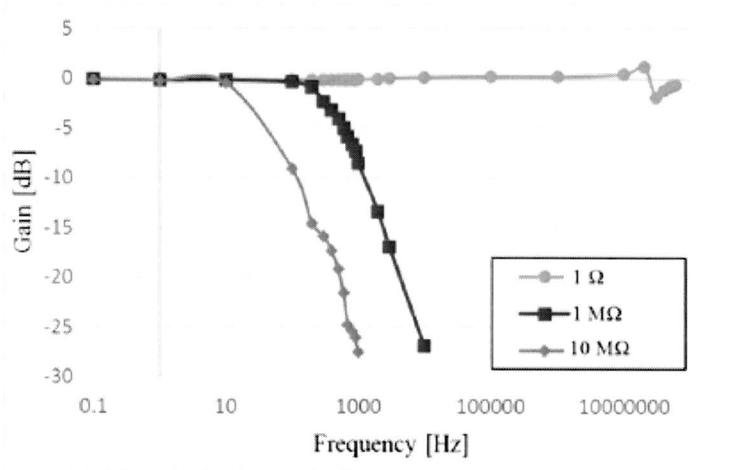

Figure 36. Response of low pass filter

High Pass Filter

In this case the PCC implant was connected to a resistive element Rf as shown in Figure 37. Also, the input resistance to the oscilloscope was taken into account. Figure 38 shows the filter response for resistors of 1 Ω, 1 MΩ and 10 MΩ. When the Rf resistance was 10 MΩ, the response obtained was that of a high pass filter with a cutoff frequency of 49.13 Hz.

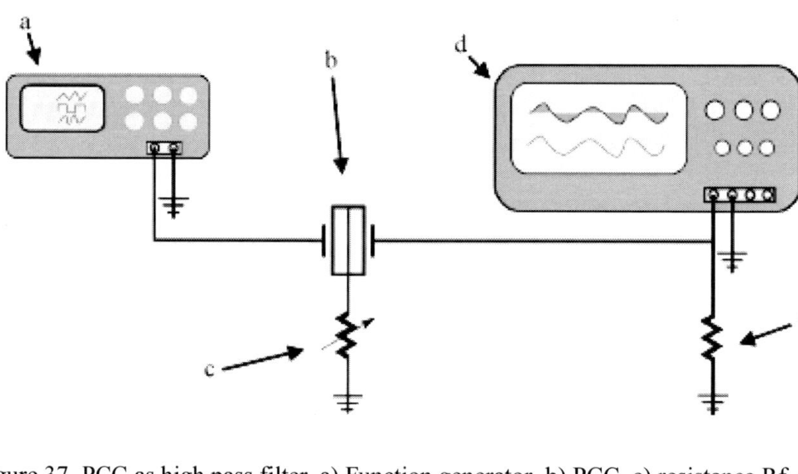

Figure 37. PCC as high pass filter. a) Function generator, b) PCC, c) resistance Rf, d) oscilloscope and e) input impedance of the oscilloscope.

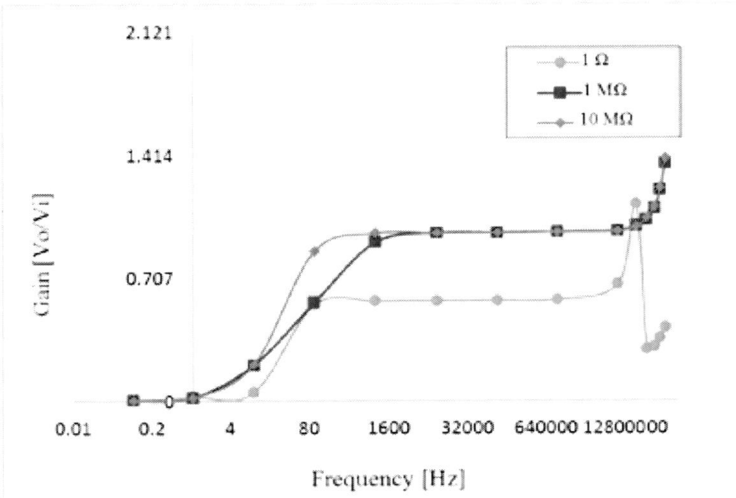

Figure 38. Response of high pass filter.

Digital Electronics

AND Gate

In the field of digital electronics, the PCC can be used to mix square waves. For this experiment, the faces of the PCC were excited with square waves at 80 kHz and 10 kHz as seen in Figure 39. A mixed signal was obtained by means of the implant as can be seen in Figure 40. An AND gate was obtained using a rectifier circuit for the resulting signal, where the faces of the ceramic work as digital inputs and the implant as output.

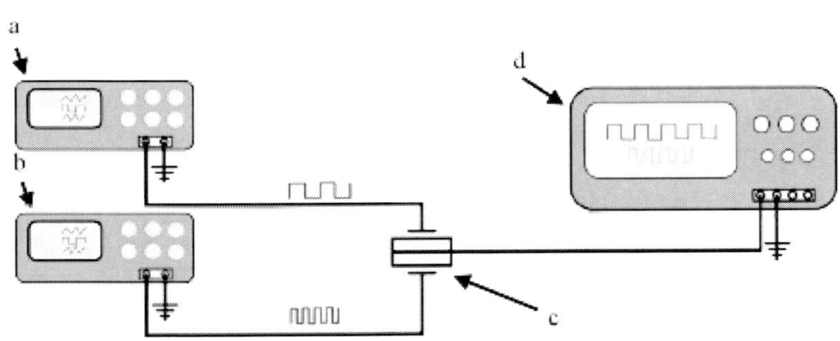

Figure 39. PCC as AND gate. (a,b) Function generators, c) PCC and d) oscilloscope.

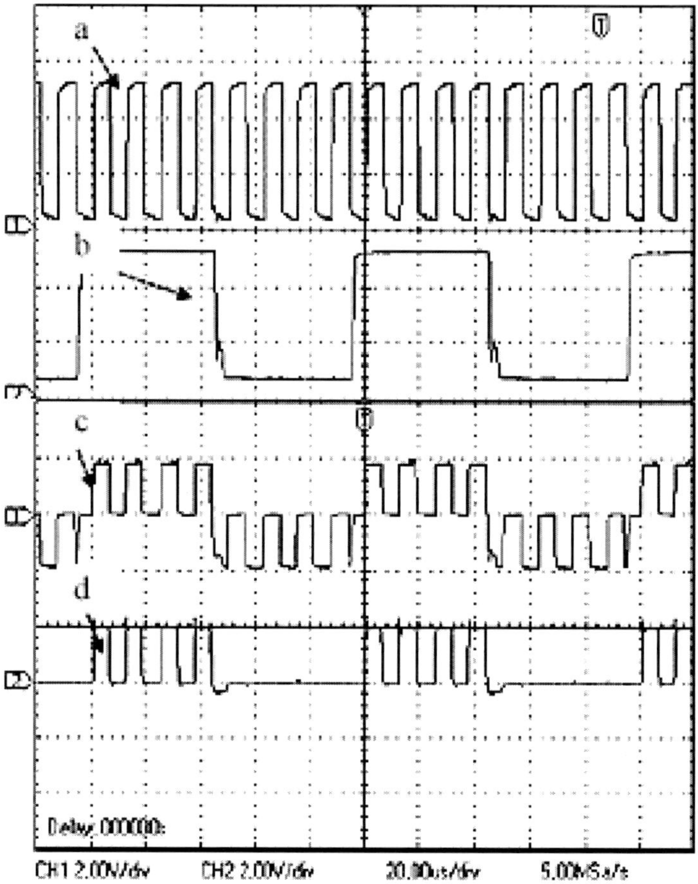

Figure 40. Response of PCC as AND gate. a) Square input wave at 80 kHz, b) square input wave at 10 kHz, c) output signal in the implant and d) AND behavior after of rectification.

Hybrid NOR Gate

In this experiment, the circuit shown above was complemented with a MOSFET transistor as shown in Figure 41. The MOSFET transistor is a semiconductor that has a high input impedance in the gate. Therefore, it is ideal for coupling the piezoelectric ceramics since the output of these presents a very high impedance and a very small output current.

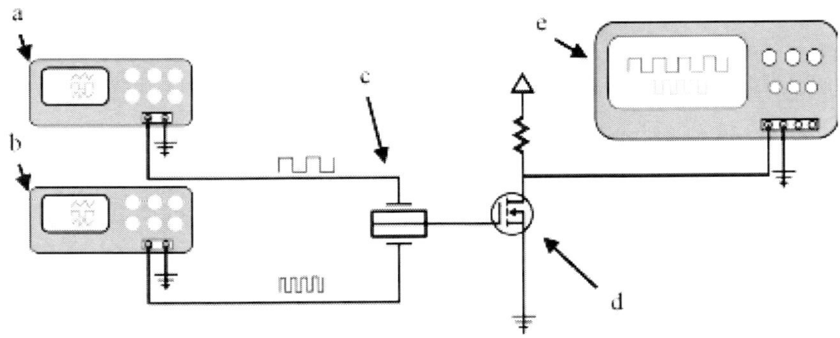

Figure 41. PCC as NOR gate. (a,b) Function generator, c) PCC, d) MOSFET and e) oscilloscope.

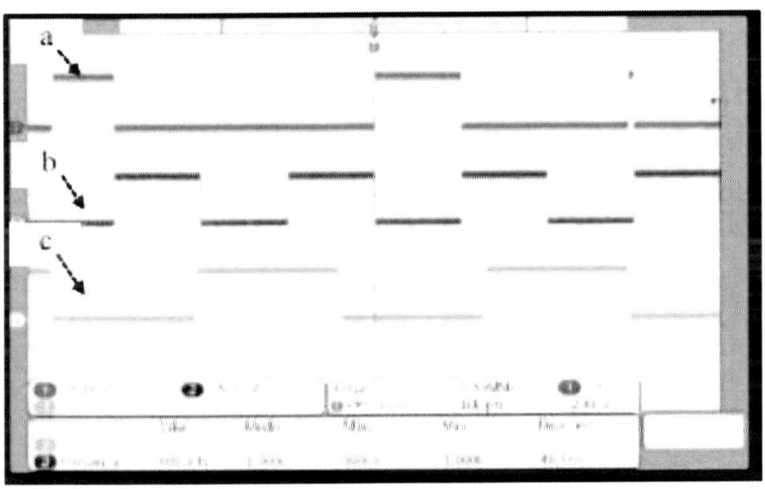

Figure 42. Response of PCC as NOR gate. a) Square input wave at 10 kHz, b) square input wave at 80 kHz, c) output signal in the implant and d) NOR behavior after of the MOSFET.

For this experiment, TTL at 80 kHz and 10 kHz frequencies were used to excite the faces of the PCC. As shown in Figure 41, the transistor works as a rectifier and inverter circuit for the signal that is obtained in the implant. The signals shown in Figure 42 correspond to the behavior of a NOR gate. The NOR gate is a universal gate with which all other gates can be created.

Mechanics

Impact Detector of an Engine

Taking advantage the piezoelectric characteristics of the PCC, it is used, in this application, as an impact sensor in an engine. The PCC was mounted on a 24 V CD motor with an analog tachometer as shown in Figure 43.

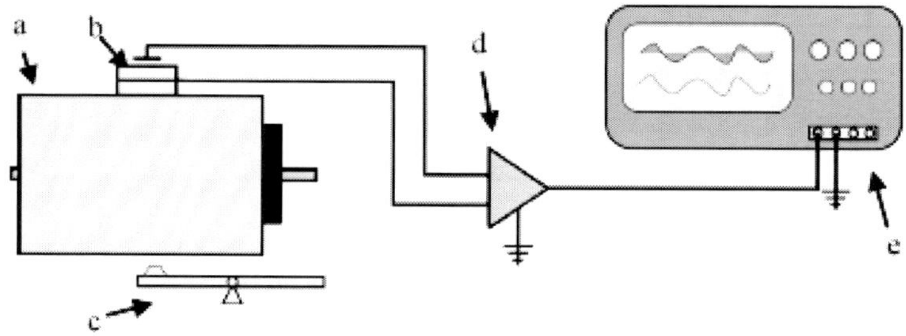

Figure 43. PCC used as vibration detector. a) CD motor with tachometer, b) PCC, c) mechanical stimulator, d) amplifier and e) oscilloscope.

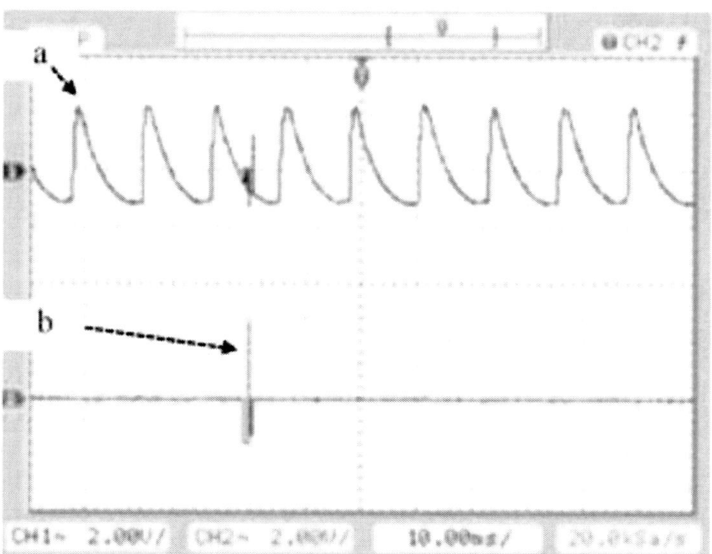

Figure 44. Response of vibration detector. a) Tachometer signal and b) PCC signal when the motor is impacted.

The CD motor was struck in a pulsating manner and controlled by a mechanism developed for this specific purpose. The electric charges generated in the PCC due to the induced vibration were amplified and recorded. The signal coming from the tachometer and the response of the PCC can be seen in Figure 44.

Piezoplethysmograph

The volume of blood in the arteries is modified according to the cardiac cycle. These volume changes can be recorded by means of sensors placed on the tip of the finger, ear lobe and other parts of the body. In this particular application, changes in blood volume were used to induce a mechanical deformation in the PCC, this way, sensing the cardiac pulse [8]. One of the faces of the ceramic was left free and was used as a sensory area where the index finger was placed as shown in Figure 45.

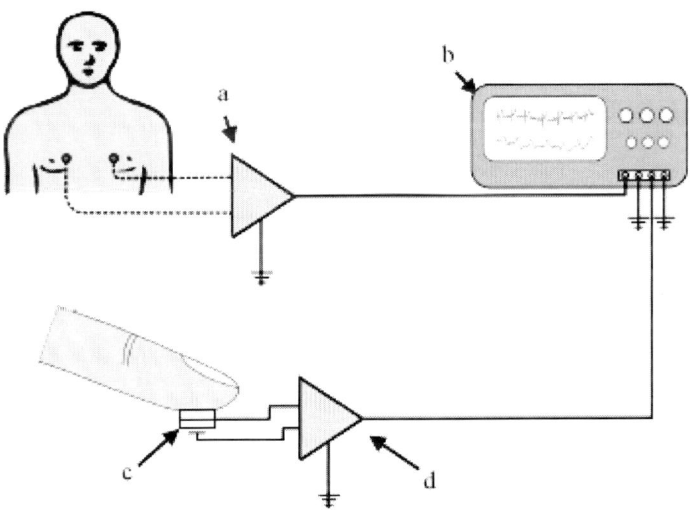

Figure 45. Piezoelectric plethysmograph. a) ECG amplifier, b) oscilloscope, c) PCC and d) amplifier.

The metallic implant was used as an electrode and allowed recording the electric charges generated in the PCC due to the change in blood volume at the tip of the finger. In order to validate this, a test subject was placed in supine position to record the electrical activity of his heart by means of an

ECG. The ECG signal was measured with the commercial device Bio Amp 100 (Axon Instruments). Simultaneously, the pulse signal obtained in the PCC was recorded. Both signals shown in Figure 46 were observed in the oscilloscope which served as a cardiac monitor.

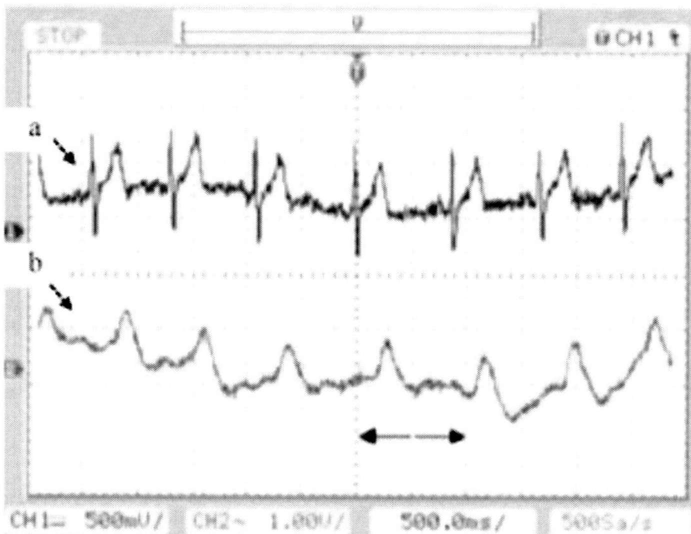

Figure 46. Response of piezoelectric plethysmograph. a) ECG signal and b) plethysmograph signal of the PCC.

Optics

Visible Light Detector

In this regard, the PCC was used as visible light detector. One face of the PCC was irradiated by a stroboscope at 3.1 Hz at a distance of 5 cm as shown in Figure 47. In order to compare the PCC response, a commercial light sensor was used at the same time. In this test, an electrical signal synchronized to the light beam was obtained through the Pt implant. The response signal of the PCC and the commercial sensor response were registered and can be seen in Figure 48.

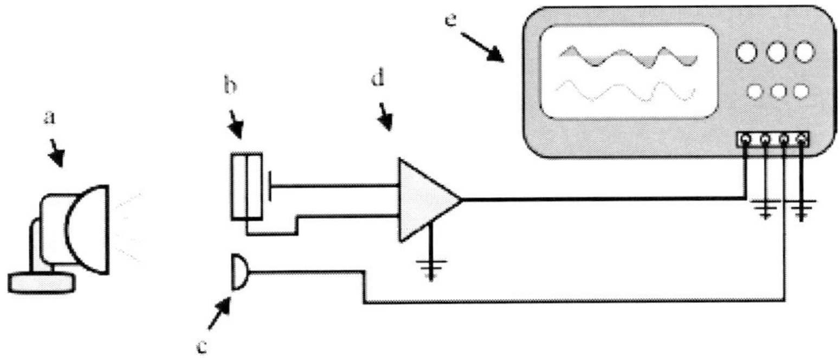

Figure 47. Visible light detector. a) Strobe light, b) PCC, c) commercial light sensor, d) amplifier and e) oscilloscope.

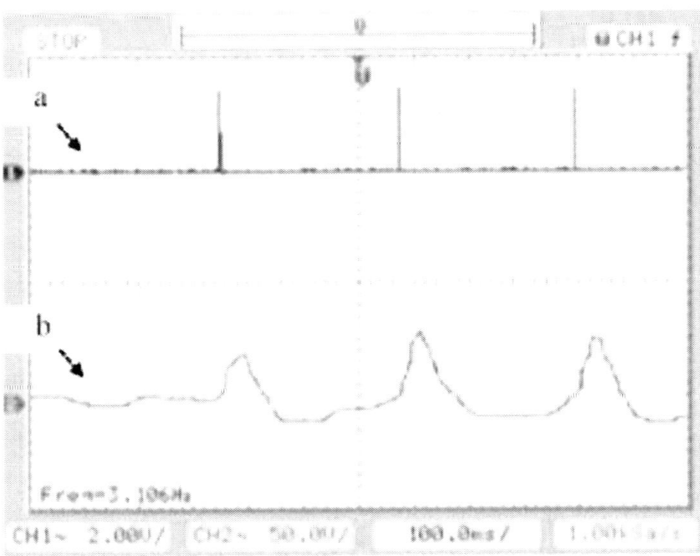

Figure 48. Response of light detector. a) Commercial sensor signal and b) PCC signal.

Laser Light Detector

In this test, an electrical response provoked by the incidence of a pulsating laser light on one face of the PCC was obtained. The laser (He-Ne) power was 10 mW/cm^2 and the pulse repetition frequency was 2.11 Hz. The experimental setup is shown in Figure 49.

A reference signal was obtained by the chopper controller in order to compare the amplified PCC signal. The response signals were registered in an oscilloscope and both signals are exhibited in Figure 50.

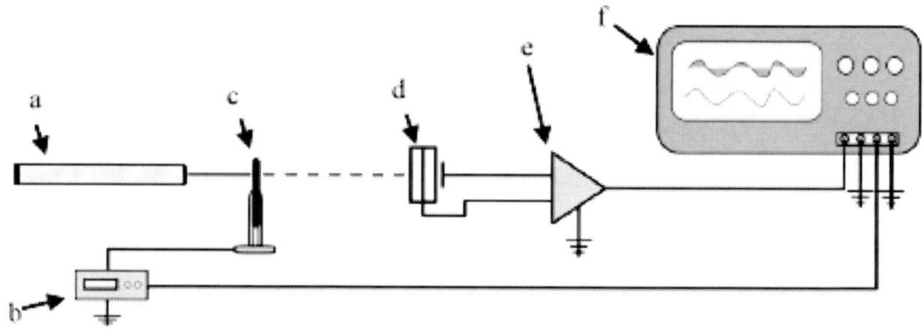

Figure 49. Laser detector. a) Laser, b) chopper controller c) chopper, d) PCC, e) amplifier and f) oscilloscope.

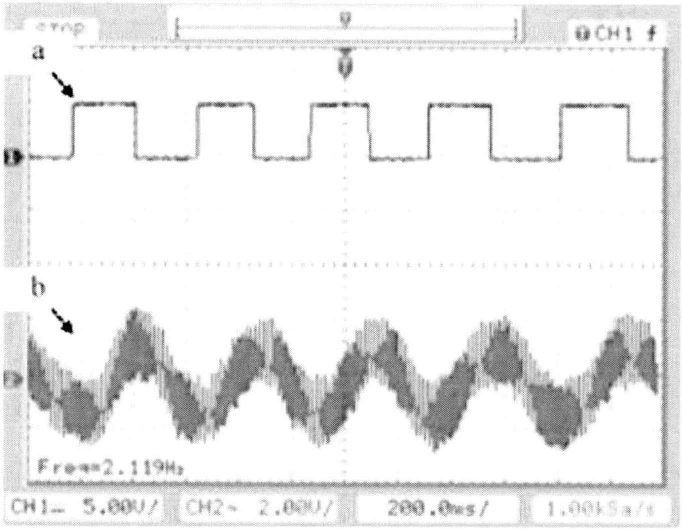

Figure 50. Response of laser detector. a) Signal of the chopper controller and b) PCC signal.

Opacity Sensor

The set shown in Figure 51 was used to obtain the photoelectric response in the PCC, which is stimulated by a two-dimensional light sweep.

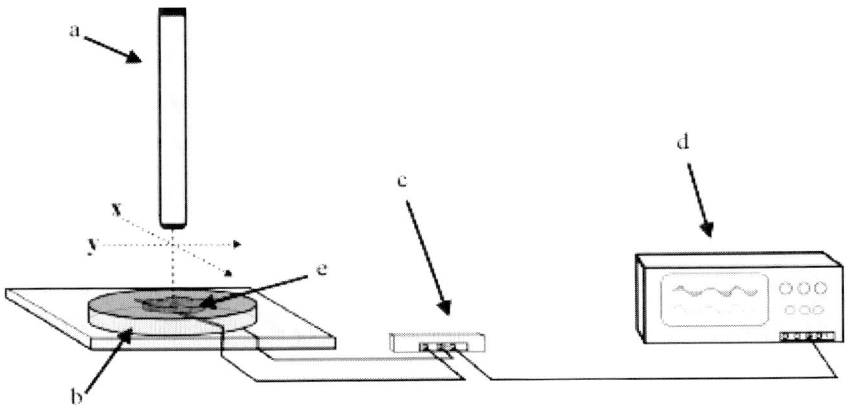

Figure 51. Experimental setup for two-dimensional sensing. a) Laser, b) PCC, c) amplifier, d) oscilloscope and e) optional sample.

Firstly, a characteristic curve of the PCC response must be obtained. Figure 52 shows photoelectric response of the PCC along the x and y axes with a resolution of 1 mm.

Once this response is known, any sample can be placed on the top of the PCC in order to know its opacity along the x and y axes.

The opacity of a tree leaf was measured. Figure 53 shows the opacity of different points of the tree leaf. The opacity decreases in the periphery of the leaf because this one does not cover all the surface of the PCC.

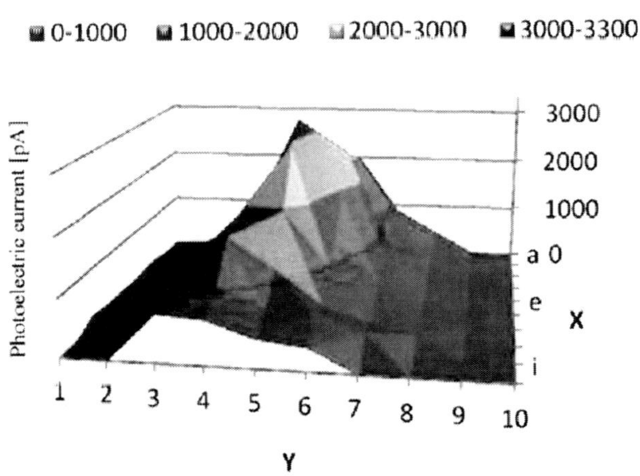

Figure 52. Response of the PCC.

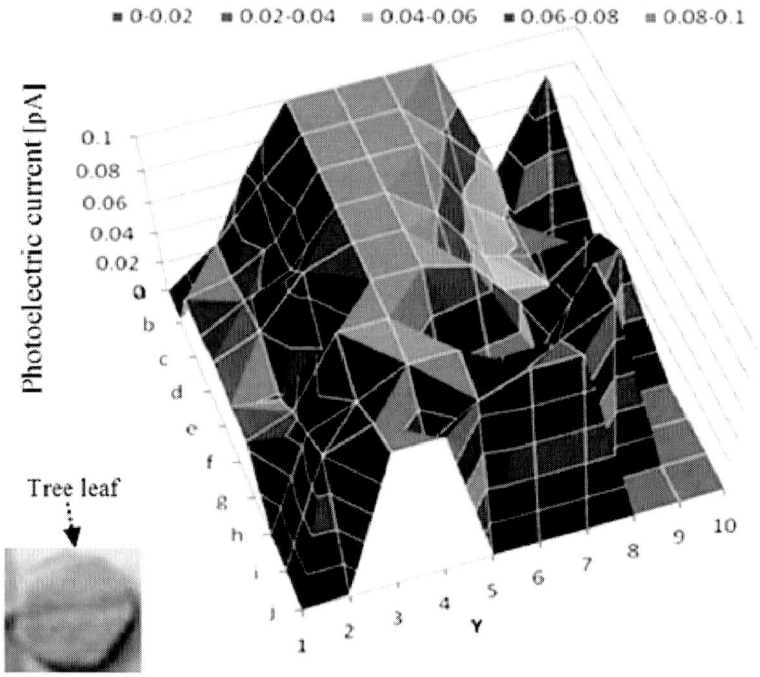

Figure 53. Opacity of the tree leaf.

Actuators

Control of the Permeability of the PCC

The implant in a PCC works as an input element that allows to modify the properties of a piezoelectric ceramic. In this regard, a porous ferroelectric ceramic with a metallic implant based on Barium Titanate and Quartz was placed as a burette seal filled with distilled water as shown in Figure 54. Later, a sinusoidal signal with a constant amplitude of 5V was applied at different frequencies (100Hz, 500Hz, 1KHz and 1MHz) in the implant and the permeability coefficient was determined [9]. The graph of Figure 55 shows the results obtained from the coefficient of permeability of the PCC at different stimulation frequencies. As can be seen in the graph, the signal connected to the implant modifies the permeability of the ceramic.

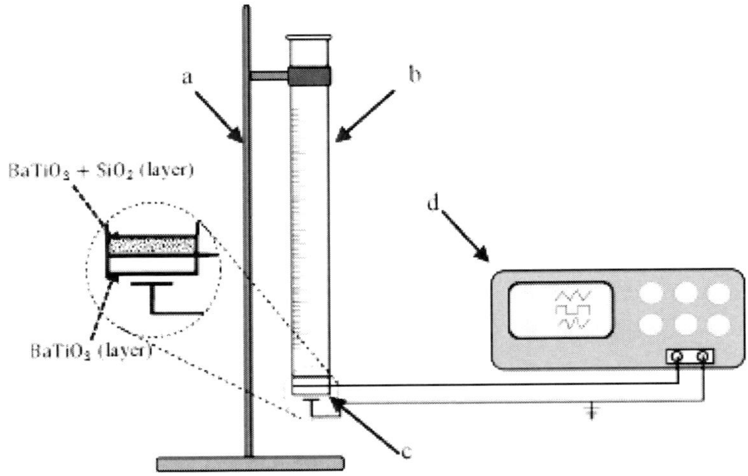

Figure 54. Experimental set up to determine the permeability coefficient. a) Universal support, b) burette, c) porous PCC and d) function generator.

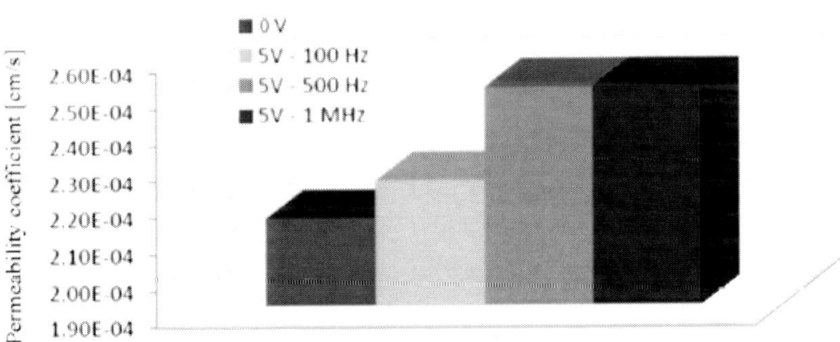

Figure 55. Permeability coefficient of porous PCC.

CONCLUSION

The piezoelectric controlled ceramic has shown a high application potential as sensor and transducer. A peculiar characteristic of the PCC is that shows a greater sensitivity in the area near the implant.

The implant works as an input or output electrode which allows that the surface of one of the faces of the ceramic to be used as a sensing area.

This feature is very useful to sense light since the face of the ceramic has direct contact with the beam.

One of the most interesting properties of the PCC and that modifies the conception of a conventional piezoelectric ceramic is the possibility to control the dielectric constant of the material. This property allows to make a frequency modulator of great utility in the field of wireless communications. Additionally, it is also possible to control the permeability of the PCC. This peculiar characteristic opens up a complete field of study to know how much the piezoelectric characteristics of a ceramic can be modified by means of a control signal.

The PCC is a device of recent application in field of piezoelectric materials and although it has demonstrated a great potential of application, it is still necessary to deepen in the phenomena involved in the behavior of the device, which will allow expanding the application possibilities of this innovative device.

REFERENCES

[1] Gonzalez-Moran, C. O.; Suaste-Gomez, E. Developed and Experimental Evidence of a Ceramic-Controlled Piezoelectric Bulk Implanted with Pt Wire Based on PLZT. *Ferroelectrics,* 2009 392, 98-106.

[2] Suaste-Gomez, E.; Flores-Cuautle, J. D. A.; Gonzalez-Moran, C. O. Opacity Sensor Based on Photovoltaic Effect of Ferroelectric PLZT Ceramic with Pt Wire Implant. *IEEE Sens. J.,* 2010 10, 1056-1060.

[3] Gonzalez-Moran, C. O.; Cruz-Orea, A.; Flores-Cuautle, J. J. A.; Minor-Martinez, A.; Elias-Vinas, D.; Suaste-Gomez, E. Ceramic-Controlled Piezoelectric Bulk Implanted with Pt Wire Based on BaTiO3 (Optical Microscopy, SEM, EDS) and PLZT (Optical Bi-dimensional Characterization). *Ferroelectrics,* 2011 423, 105-110.

[4] Suaste-Gomez, E.; Gonzalez-Moran, C. O.; Teran-Jimenez, O.; Rodriguez-Montoya, C. J. Opto-Thermal Response of Ceramic

Controlled Piezoelectric Based on PLZT Implanted with Platinum. *Ferroelectrics,* 2014 467, 158-164.

[5] Jaffe, B.; Cook, W. R.; Jaffe, H. L., *Piezoelectric ceramics*. London; New York: Academic Press, 1971.

[6] *Key paper in physics - Piezoelectricity*. New York: American Institute of Physics, 1991.

[7] Uchino, K., *Advanced piezoelectric materials: science and technology*. Cambridge: Woodhead Publishing Ltd, 2010.

[8] Gonzalez-Moran, C. O.; Flores-Cuautle, J. J. A.; Suaste-Gomez, E. A Piezoelectric Plethysmograph Sensor Based on a Pt Wire Implanted Lead Lanthanum Zirconate Titanate Bulk Ceramic. *Sensors-Basel,* 2010 10, 7146-7156.

[9] Acuautla-Meneses, M. I.; Flores-Cuautle, J. J. A.; Suaste-Gomez, E. Permeability Behavior of Porous Piezoelectric Ceramics Prepared in Layers in Response at AC Voltage on Electrode Implanted. *Ferroelectrics,* 2011 423, 111-115.

In: Advances in Engineering Research
Editor: Victoria M. Petrova
ISBN: 978-1-53614-218-1

Chapter 4

PASSIVE MAGNETIC BEARINGS

Milad AlizadehTir[1,*], Fabrizio Marignetti[1] and Sorin Ioan Deaconu[2]
[1]Electrical and Information Engineering Department (DAEIMI), University of Cassino and Southern Lazio, Cassino (FR), Italy
[2]Electrical Engineering and Industrial Informatics Department of Politecnica University TIMIŞOARA, România

ABSTRACT

The analysis of the behaviour of Passive Magnetic Bearing in order to achieve an acceptable magnetic force and stiffness is an interesting topic for rotating systems. Numerical analysis, which is an effective method to investigate the structural parameters of PMB, is applied using Finite Element Method to the two-dimensional model of Passive Magnetic Bearing. Numerical analysis is benefecial in predicting the performances of the bearing versus different changes in the dimensions of the PMB. An optimization through Genetic Algorithms is then performed. The data gathered from the numerical analysis are therefore transferred to the Genetic Algorithm to facilitate the definition of the fitness and penalty functions which will help a faster convergence to the

* Corresponding Author Email: milad@unicas.it.

objective function. Providing a method to improve the magnetic force and consequently the magnetic stiffness of Passive Magnetic Bearings is an important purpose of this chapter. Also, in order to compare Passive Magnetic Bearing with different dimensions, the force to cost ratio is proposed as an index considering magnetic force and economical factors. The Genetic Algorithm is a stochastic optimization method which can be applied in order to reach the best dimensions according to the considered objective functions.

Keywords: force-expenses ratio, genetic algorithm, numerical analysis, passive magnetic bearing

INTRODUCTION

Energy saving is a key issue for industries, especially for those using rotational systems [1-3]. Friction, power loss, limitations in rotation speed, periodic lubrication and maintenance are the most important disadvantages of mechanical ball bearings. Magnetic levitation was the object of Earnshaw's theory, which was published in 1842 and since then, it has been having a strong impact upon industries [3-8]. Magnetic bearings help solving industrial problems in some applications like aerospace, clean room atmospheres, severe environments e.g., high rotational speeds like turbo-molecular pumps, turbo-compressors and flywheels for energy storage [9-10].

Magnetic bearings are contactless bearings which achieve various technical advantages such as: no wearing and long life span, extreme reliability, frictionless nature, lubrication free operation, low losses, low maintenance cost, low acoustic noise, zero moment of resistance and applicability to extreme operating conditions, e.g., for very high rotational speeds, low temperature, radiation and vacuum environments. These advantages are valuable in comparison to conventional mechanical ball bearings [1-8], [11].

Stable magnetic suspension was realized by F. Holmes and J. Beams in the centrifuges in the mid-1930s. The early studies concerning Passive

Magnetic Bearings (PMBs) with permanent magnets have been done by Yonnet using modern rare earth permanent magnets [12-13].

The rare-earth permanent magnets allow using repulsive force in order to achieve magnetic levittation.

According to the structure of the magnetic bearing and to the strategies for controlling the position of the rings, magnetic bearings are classified into PMB, Active Magnetic Bearing (AMB) and Hybrid Magnetic Bearing (HMB). Another classification can be made according to the magnetization direction of the rings, which can be Axial or Radial. There are even other ways to categorizing magnetic bearings which is not the purpose of this chapter [9], [14-16].

AMB includes magnets, coils and a power supply together with a control circuit to detect the misalignment and set the appropriate current to produce the relevant magnetic field. This structure is suitable for high-precision applications, however due to its high cost, it is not economically justifiable for all industry applications. HMB uses permanent magnets assisted by electromagnets to achieve a pilot and ready to use magnetic field together with a controlled magnetic field. HMB needs a control circuit and a supply source. Despite their precision, the complexity of AMB and HMB and their high cost have increased the attention towards PMB. Therefore, a comprehensive investigation on PMB is needed [17-19].

The Earnshaw's theorem states that perfect (i.e., in all degrees of freedom) passive magnetic levitation is impossible and consequently at least one degree of freedom has to be mechanically, actively or electro-dynamically fixed in rigid bodies with five degrees of freedom (namely five-DOF systems) [20-21]. In comparison with other types of magnetic bearings, PMB can tolerate weaker magnetic forces and consequently achieves lower magnetic consistency. For this reason, the characteristics of PMB should be investigated in order to solve this deficit. In the following, a brief assessment is done on these characteristics using the Finite Element Method (FEM). PMBs are usually manufactured with ring permanent magnets in order to levitate a rotor in a rotational system. The simplest structure of PMB includes two concentric rings separated by a cylindrical air-gap or two rings with the same dimensions separated by a plane air-gap. PMBs can be magnetized in the axial or in the radial directions.

In the following, the most important characteristics of PMB such as magnetic force and magnetic stiffness and the relation between them are explained. Numerical analysis is performed using 2-Dimensioanal FEM (2-D FEM) [8], in order to define fitness and penalty functions to build a Genetic Algorithm (GA) - based optimization technique. Force to cost ratio is proposed in order to perform a comparison. Finally, GA is used to suggest a way to achieve the best dimensions that maximize the force with the lowest magnets volume.

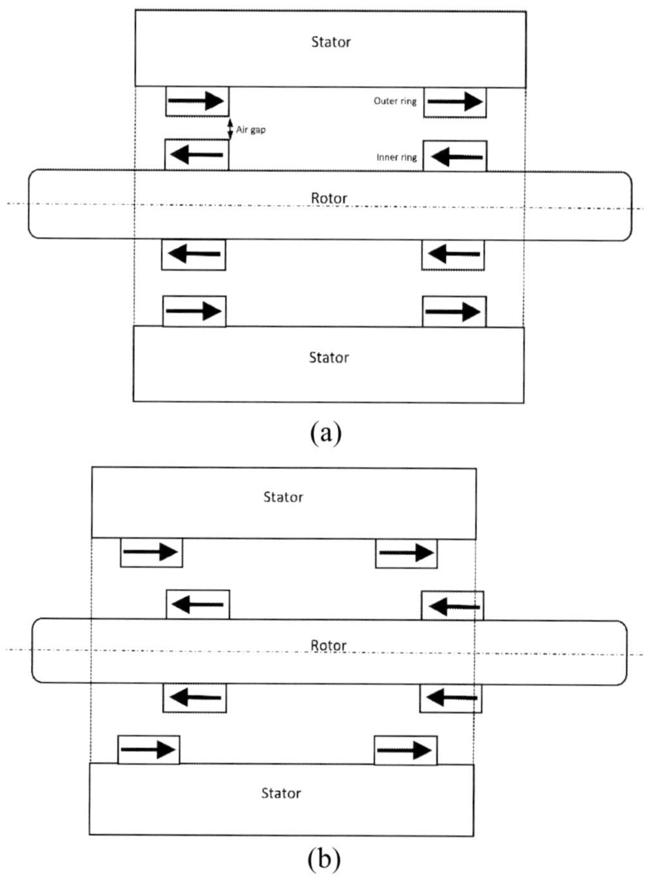

Figure 1. Front view of PMB installed on the rotor and stator, (a) stable position, (b) misaligned situation.

MAGNETIC FORCE AND STIFFNESS

In the repulsive type of magnetic bearing systems, the magnetic levitation of the rotor is realized using the interaction forces between the magnetic rings, which are coupled to the stator and to the rotor as in figure 1. The most important characteristics of the PMB structure are the magnetic force (F) and mechanical stiffness (K) they produce. The resultant force due to the interaction between the magnetic fields produced by rings, generate the magnetic levitation, which counteracts the weight of the load, rotor and shaft to be levitated. Figure 1(a) shows the stable position of rotor and stator in which the axial displacement is zero.

If a displacement happens as shown in figure 1(b), the magnetic stiffness is responsible for returning the system into the stable point.

The relation between the magnetic force and the stiffness can be the same as the relation between the mechanical force and the stiffness in the spring, therefore magnetic force and magnetic stiffness can be expressed as:

$$\vec{F} = F_x \cdot \vec{x} + F_y \cdot \vec{y} + F_z \cdot \vec{z} \tag{1}$$

$$\vec{K} = -\left(\left(\frac{dF_x}{dx} . \vec{x} \right) + \left(\frac{dF_y}{dy} . \vec{y} \right) + \left(\frac{dF_z}{dz} . \vec{z} \right) \right) \tag{2}$$

$$\vec{K} = -(K_x \cdot \vec{x} + K_y \cdot \vec{y} + K_z \cdot \vec{z})$$

Figure 2(a) shows a mechanical ball bearing in a rotating device and Figure 2(b) shows the simplest conventional structure of PMB which consists of two concentric rings separated by a cylindrical air-gap. Earnshaw's theorem states that the sum of the magnetic stiffness in the PMB must be equal to zero, i.e.:

$$K_x + K_y + K_z = 0 \tag{3}$$

$$K_x = K_y = K_z$$

Because of the axial symmetry, Eq.(3) becomes:

$$K_z = -2K_r \tag{4}$$

The ratio between axial and radial stiffness is always - 2 for pure permanent magnetic configurations with rotational symmetry [15]. If K_z is known, then the stiffness of the magnetic bearing is perfectly determined. To achieve stable configuration of any bearing, the stiffness needs to be positive, although equation (4) indicates that positive values of K_r result into negative values of K_z, so there is a very strong coupling between axial and radial stiffness of cylindrical radial magnetic bearing [22-23].

Radial permanent magnet bearings magnetized in the axial direction are unstable in this direction so to guarantee stability of radial permanent magnet bearing along the axis of rotation, two thrust bearings are necessary. As it is seen in figure 3, a mechanical holder is attached to the outer ring in order to satisfy Earnshaw's theory.

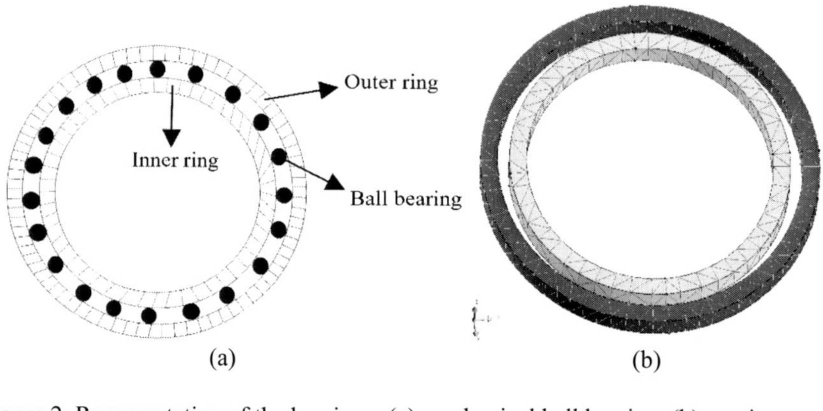

Figure 2. Representation of the bearings: (a) mechanical ball bearing, (b) passive magnetic bearing.

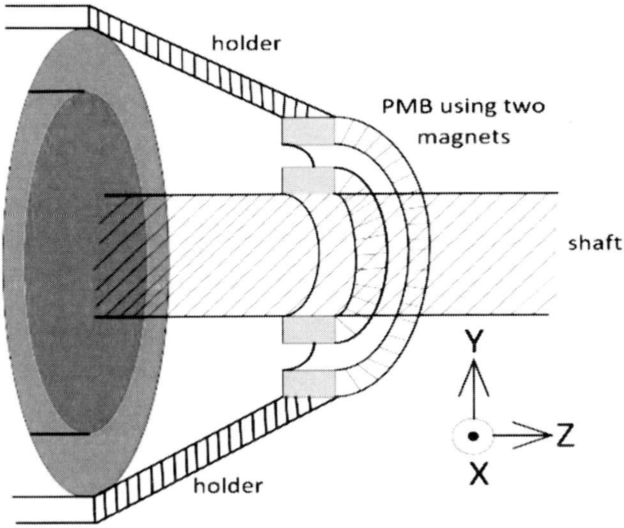

Figure 3. Passive Magnetic Bearing.

Table 1. Specifications of the PMB

Symbol	Quantity	Value	Symbol	Quantity	Value
r_{o_out}	Outer radius of the outer ring	28 mm	h_o	Outer ring height	3 mm
r_{0_in}	Inner radius of the outer ring	25 mm	h_i	Inner ring height	3 mm
r_{i_out}	Outer radius of the inner ring	24.8 mm	J	Magnetic polarization vector	1.1 T
r_{i_in}	Inner radius of the inner ring	21.8 mm			

In the PMB structure, the outer magnet is fixed while the inner is attached to the rotor. The interaction between the magnetic field of the two magnets provides magnetic levitation and allows rotation of the shaft.

In order to investigate the behavior of PMB, the parameters which describe the geometry of the PMB structure in figure 3, are shown in figure 4. Table 1 explains these parameters together with their considered values.

As it is presented in figure 4, the magnets are radially centered and their magnetization is considered uniform and parallel to axis, but in opposite directions, to provide the necessary interaction.

FINITE ELEMENT SIMULATION

There are different methods to investigate the axial force and stiffness in magnetic bearing structures, i.e., analytical or semi-analytical methods. Each method has some positive aspects and some drawbacks. As an example, elliptic integrals are used in the analytical method, which result in a complex calculation formula. The flux linkage can be another important factor in the magnetic bearing structure which must be investigated by the numerical method. Computational cost is an interesting factor in evaluating the analytical methods.

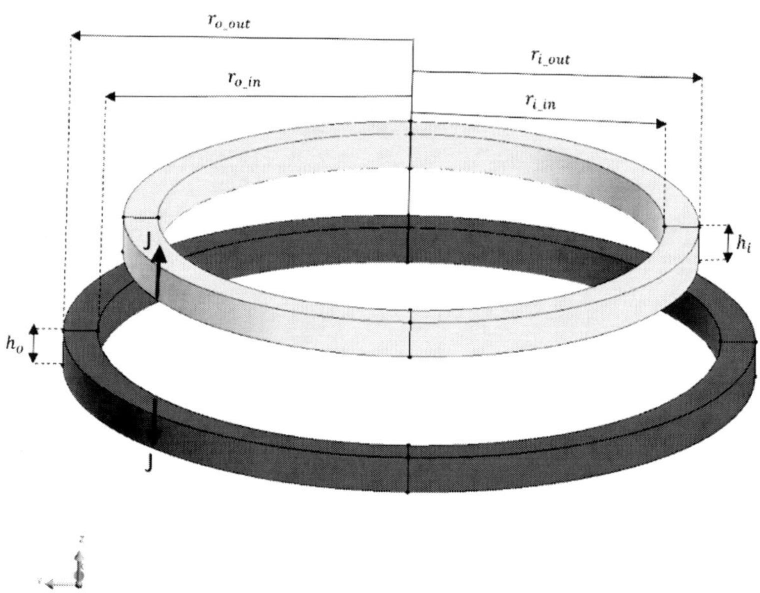

Figure 4. Main geometrical dimensions of the PMB.

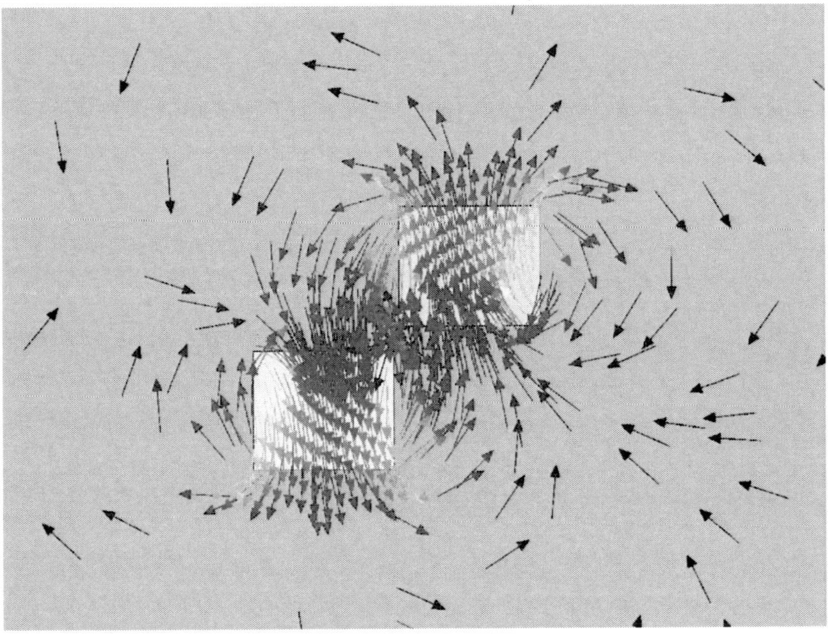

Figure 5. 2D-FEM simulation of PMB together with axial magnetization.

In the absence of radial displacement, as can be seen in figure 4, the structure of PMB has an intrinsic symmetry around the vertical axis. This is used in the numerical analysis in order to simplify the model and reduce the computational time

Finite Element Method (FEM) performs a precise analysis of magnetic devices taking into account geometric details, magnetic non-linearity and also leakage flux of PMB. FEM divides a magnetic device into a finite number of elements and numerically solves the Maxwell equations for each section, then applies the superposition principle. Due to the annular structure and consideration of the symmetrical properties of the PMB structure, 2D-FEM can be applied instead of the Three-Dimensional (3D) simulation.

A simple simulation of a PMB with axial magnetization, according to the dimensions given in Table 1 is presented in figure 5.

NUMERICAL ANALYSIS

Numerical analysis is performed by considering the most important parameters of PMB i.e., air-gap length, width and height of the magnetic rings.

The results of the numerical analysis can be useful to constitute the initial population of the Genetic Algorithm (GA) which will be detailed in Section Genetic Algorithm. A first guess of the best dimensions is predicted by means of the numerical analysis. The following criterion is adopted:

$$Force-cost\ \ ratio = \frac{Maximum\ amplitude\ of\ \ the\ \ magnetic\ \ force}{Volume\ of\ \ used\ \ magnet} \tag{5}$$

Force-cost ratio considering the characteristics of PMB like magnetic force and the economic matters like the cost of the magnets may be used as a performance index to compare PMBs with different dimensions.

Investigation the Air-Gap Width

The first parameter to vary in the numerical analysis is the air-gap width between two magnets, i.e., between r_{i_out} and r_{o_in}. The investigation of the air-gap length is performed according to the dimensions in Table 1.

When the axial displacement of the inner magnet, which is coupled to the shaft, is zero and it is in the balance point between the magnets, the magnetic force should be zero and according to the Eq. 2, the magnetic stiffness should be maximum. The result of the numerical analysis for the air-gap dimension in figure 6 and 7 show the same results. Removing the magnets from each other, the force becomes stronger to return the inner magnet into its balance position. This return procedure will be forced by the magnetic stiffness which is responsible for keeping the balance position.

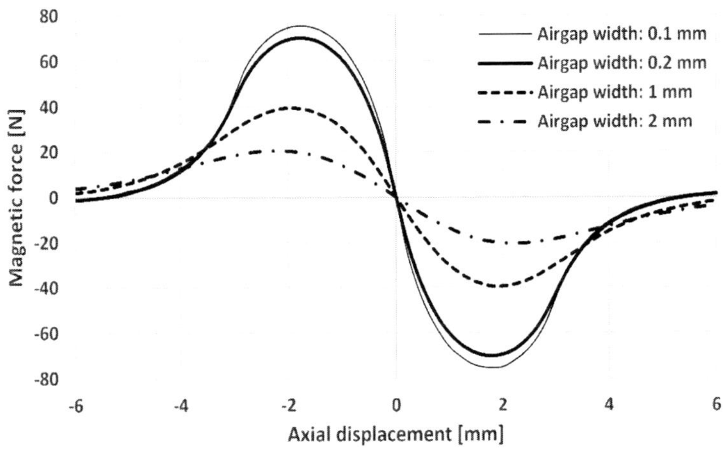

Figure 6. Magnetic force versus axial displacement according to air-gap dimension.

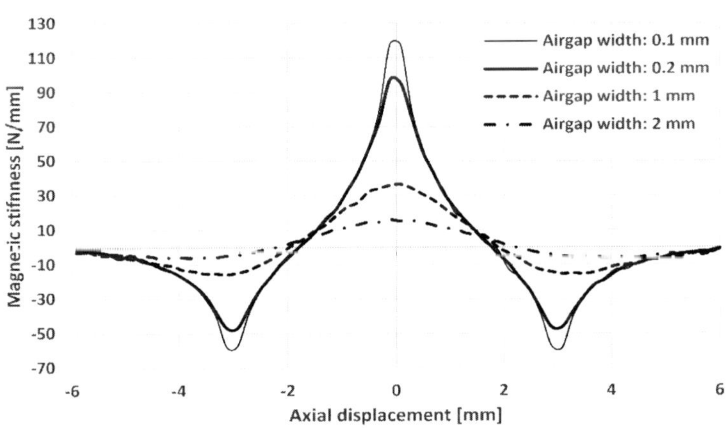

Figure 7. Axial magnetic stiffness versus axial displacement of the inner ring.

According to the figures 6 and 7, it can be concluded that smaller dimension of the air-gap will result into a stronger magnetic force and consequently stronger magnetic stiffness. On the other hand, mechanical restrictions compel the selection of 0.2 mm as the air-gap width, in order to avoid excessive friction and wearing.

Investigation of the Heights of the Magnets

In order to investigate the effect of the height of the magnets, it is assumed that the height of the outer ring is fixed and equal to 3 mm while the height of the inner ring is changed. The results of this section are presented in figure 8(a). It is concluded that when the height of both magnets are the same, the maximum amplitude of the magnetic force is achieved.

Another step of the numerical analysis is done keeping the same height for both of the magnets as shown in figure 8(b). It can be concluded that increasing the height of the magnets over 5 mm, is not effective to improve the magnetic force. A comparison is shown in figure 8(c) to show a precise presentation of the maximum amplitude of the magnetic force against different heights of the magnets. In order to compare the results more precisely, the force-cost ratio is applied.

Selecting the best dimensions of PMB depends on the magnetic force that PMB should tolerate. What is considered here is the maximum force during the investigation.

Table 2 shows that increasing the height does not improve the amplitude of the magnetic force. While the maximum amplitude of the force is achieved for the height equal to 6 mm, the difference between the maximum amplitudes the force of PMBs whose inner ring heights are 5 and 6 mm, is only 1.58% and therefore the force-cost ratio is greater for the height of 5 mm.

So considering the force-cost ratio which includes the cost of the magnets, 5 mm is selected as the best height.

Table 2. Comparison according to the force-expenses ratio for different height of the magnets

Height of magnets (mm)	Maximum force amplitude (N)	Volume of the used magnets (mm^3)	Force-expenses ratio (N/mm^3)
5	84.48	4693.54	0.017999
6	85.84	5632.25	0.015241
7	85.54	6570.96	0.013018
8	84.17	7509.66	0.011208
9	82.55	8448.37	0.009771

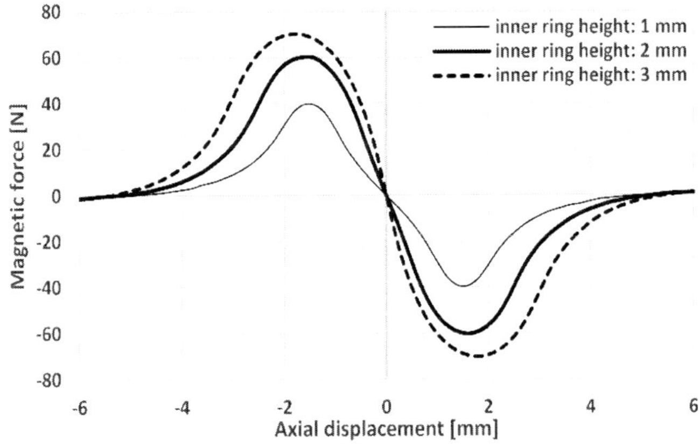

(a)

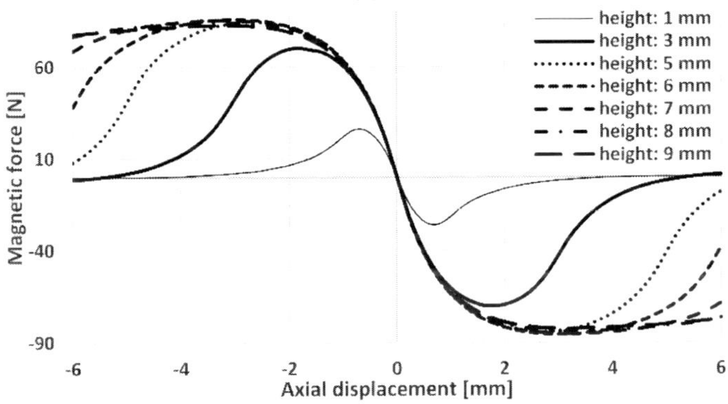

(b)

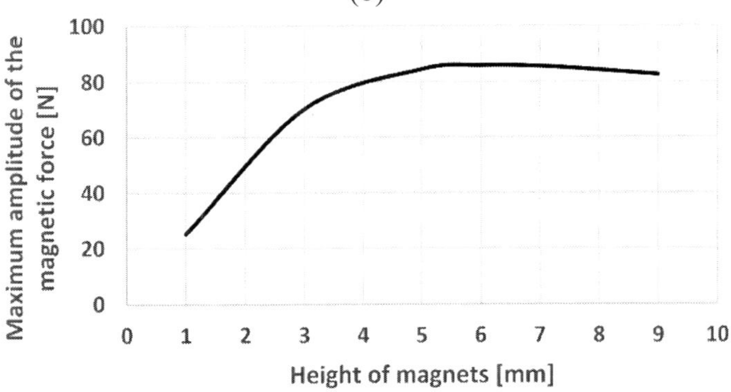

(c)

Figure 8. Magnetic force versus different height of the magnets, (a) different height for the rings, (b) the same height for the rings, (c) comparison of the maximum force amplitude.

Investigation of the Magnets Width

The final parameter which can be varied to investigate the behavior of PMB is the width of the magnets. The dimensions achieved up to now will be used in this section. The dimensions are in Table 3 while the width of the magnets are $w_o = r_{o\_out} - r_{o\_in}$ for the outer ring and $w_i = r_{i\_out} - r_{i\_in}$ for the inner ring.

Considering a fixed value for the width of the outer ring, the magnetic force is investigated versus different widths of the inner ring. The results are presented in figure 9 and it is concluded that maximum amplitude of the force will be achieved when two rings have the same width.

Table 3. PMB dimensions in order to investigate the width of the rings

Quantity	Value	Quantity	Value
r_{o_out}	28 mm	r_{i_in}	21.8 mm
r_{o_in}	25 mm	h_o	5 mm
r_{i_out}	24.8 mm	h_i	5 mm

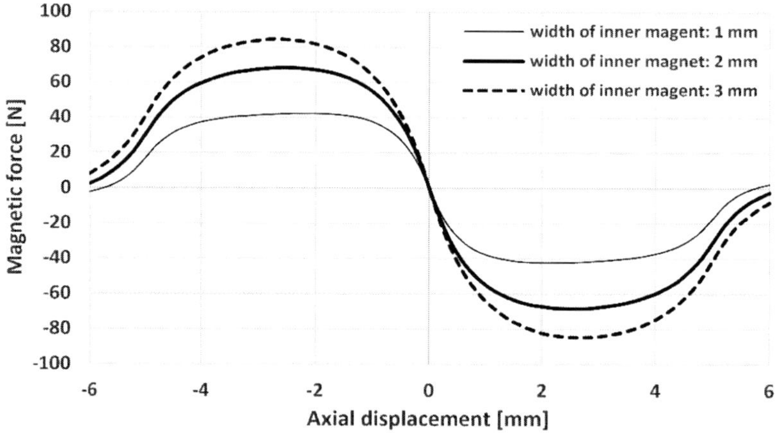

Figure 9. Force-expenses ratio versus axial displacement of the inner ring according to different widths of the inner ring.

However, it is clear that a larger magnet will improve the magnetic field and consequently the magnetic force but it should also be considered that the flux lines close their path and use all parts of the magnetic structure. As it is shown in figure 10, if the width of one ring be bigger than the width of other ring, the flux line of the ring with smaller width will close its path using some parts of the bigger ring and other parts of the magnetic ring will be useless, so it is more effective to select the same width.

Figure 11 presents the results of the investigation related to the width of the rings. Also, r_{o_in} is kept fixed equal to 25 mm and according to the results of the air-gap section, air-gap should be 0.2 mm so r_{i_out} will be 24.8 mm. This assumption is considered in order to avoid increasing the volume of the used magnet which will result in stronger magnetic field while the comparison should be done in a justify situation.

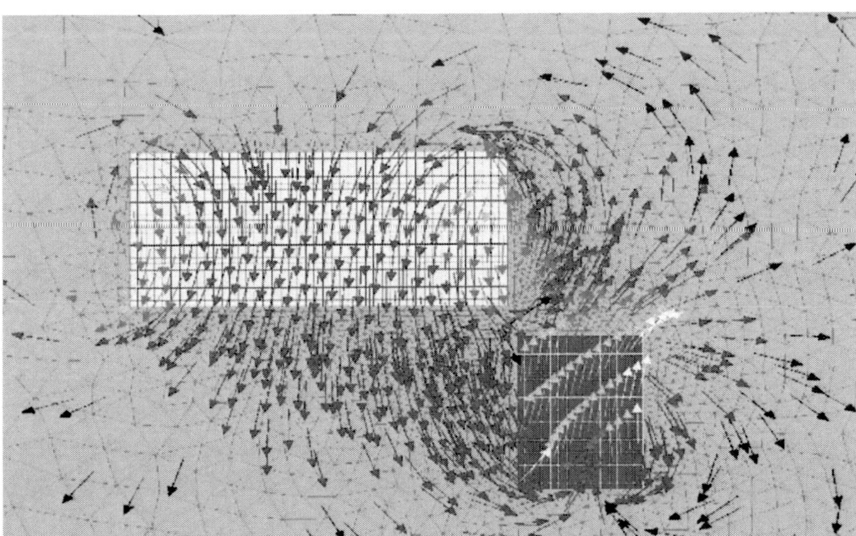

Figure 10. Presentation of the flux lines.

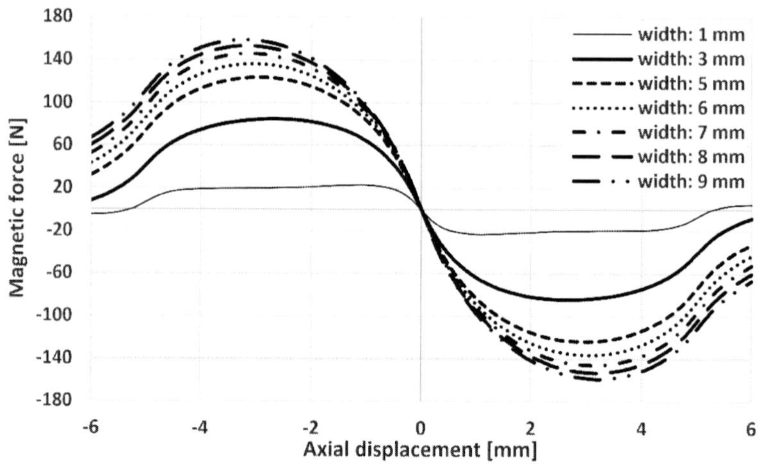

Figure 11. Force diagram versus axial displacement according to different widths of the rings.

In this step of the numerical analysis, according to the results which are presented in Table 4, it is concluded that the magnetic force amplitude will improve by increasing the width of the magnets. While on the other hand, as the result of increasing the width of the magnets, force-to-cost ratio will decrease so it is important to select a width based on a trade-off between the magnetic force and the force-to-cost ratio.

It is clear that using more volume of the magnets and consequently reaching weaker force-to-cost ratio won't be lucrative which is happing severely for the widths bigger than 6 mm, for example, the force-to-cost ratios are 0.014504 N/mm^3 and 0.0133 N/mm^3 for 6 and 7 mm, respectively. As the result, 7, 8 and 9 mm are eliminated from investigation due to their low force-to-cost ratio.

However, the force-to-cost ratio shouldn't be smaller than others; the force amplitude shouldn't be small too, so 4 mm will be eliminated due to its weak force amplitude. Magnetic forces are 136.15 and 123.78 for 5 and 6 mm respectively. Their difference will be 9.08% which is remarkable so 6 mm will be selected as the width for the magnets.

Table 4. Comparison according to the force-expenses ratio for different widths of the magnets

Width of magnets (mm)	Maximum force amplitude (N)	Volume of the used magnet (mm^3)	Force-to-cost ratio (N/mm^3)
4	106.95	6258.05	0.017089
5	123.78	7822.57	0.015823
6	136.15	9387.1	0.014504
7	145.71	10951.59	0.0133
8	153.08	12516.16	0.01223
9	158.61	14080.62	0.011264

Table 5. Result of numerical analysis

Quantity	Value	Quantity	Value
r_{o_out}	31 mm	r_{i_in}	18.8 mm
r_{o_in}	25 mm	h_o	5 mm
r_{i_out}	24.8 mm	h_i	5 mm

Result of Numerical Analysis

The air-gap width should be as small as possible, considering mechanical restrictions. The height and the width of the magnets are selected based on force-to-cost ratio and magnetic force. The results of the numerical analysis are presented in Table 5. The force diagram and the magnetic stiffness relevant to the structure using the aforementioned dimensions are shown in Figure 12 (a) and (b) respectively.

Genetic Algorithm

The structure of PMB and the dimensions of the rings must be selected according to the force amplitude and to the mechanical stiffness needed to keep PMB in a stable position with its load. Based on the numerical analysis and in order to reach a more precise sizing, an optimization method is applied according to the magnetic force and force-to-cost ratio as the objective functions, i.e., the target of the optimization. As optimization of the PMB is not a linearly constrained multi-variable problem, conventional optimization methods like the enumerated approach aren't effective. The Genetic is a very popular stochastic technique that has shown promising results and therefore is widely used for solving the optimization problems for electromagnetic devices. This technique is based on the principles of natural evolution and genetic reproduction according to Darwin's principle "survival of the fittest".

In the genetic algorithm technique, each solution is represented as a fixed-length bit string also called a chromosome. Each position in the bit string signifies a particular feature of an individual, and the value stored in that position represents how that feature is expressed in the solution. There are some operators which operate on the chromosomes in order to produce better breed.

In the topology of the genetic algorithm, definition the best objective, penalty and fitness functions is vital and important to achieve fast convergence. The fitness function is an index allowing assessing the level of accordance between each chromosome with the objective function. During each step, each chromosome with lower fitness level will be eliminated.

The results of the numerical analysis help definition the best fitness and penalty functions, because these results can help to predict the behavior of the PMB so the functions can be defined to be more effective in accelerating the convergence. The first generation, which is called "zero generation," is produced randomly and each solution is evaluated through a fitness value in order to investigate its possibility to survive. A new population is generated by applying genetic operators like crossover and mutation. Figure 13 shows the flowchart of GA used in the optimization of the PMB structure.

The most important geometric parameters of PMB are: air-gap, width of the magnets and height of the magnets. The objective function is defined to maximize the force-to-cost ratio and is evaluated when the amplitude of the magnetic force is in the desired range.

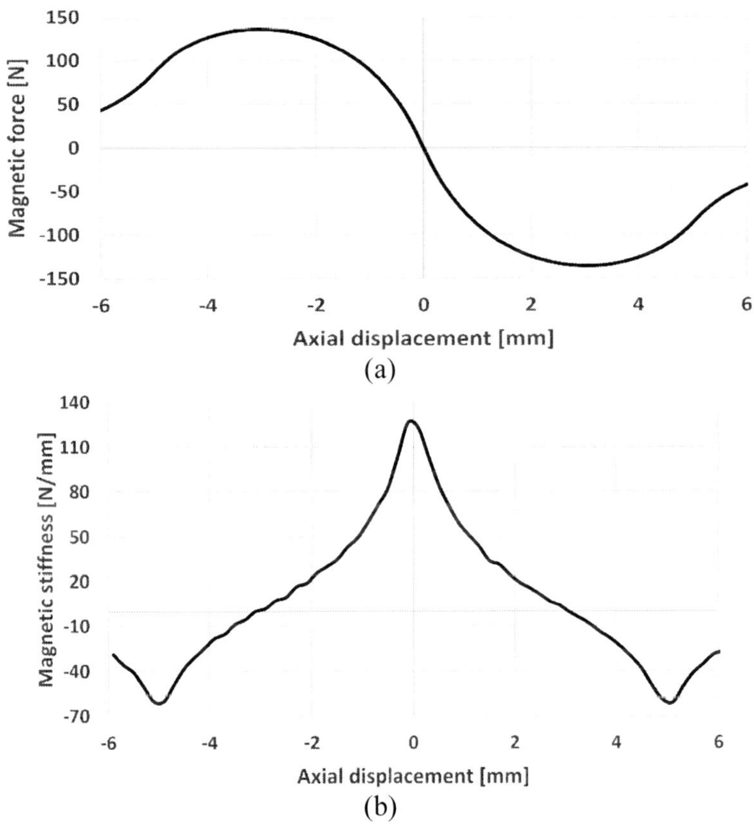

Figure 12. Characteristics of the best structure as the result of numerical analysis, (a) magnetic force diagram, (b) magnetic stiffness diagram.

Possible objective functions focus either on the magnetic force amplitude or on the magnetic stiffness or even on the force-to-cost ratio. What is considered as the objective function for the PMB is the maximum force-to-cost ratio in the defined domains for the magnetic force, height and width of the magnets. Table 6 shows the objective function of the GA.

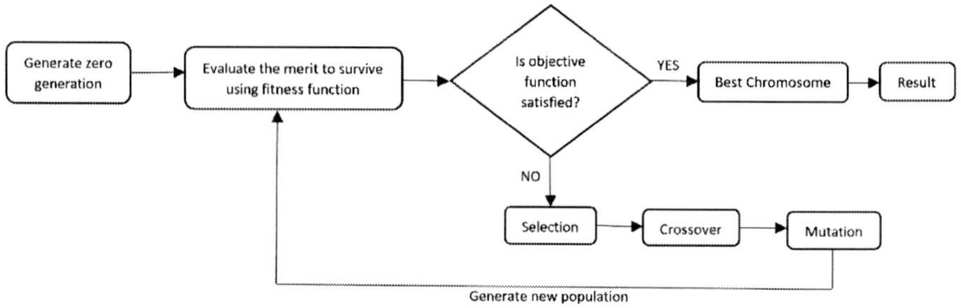

Figure 13. Flowchart of Genetic Algorithm.

The results of the optimization achieved by the GA is presented in Table 7. Figure 14 shows the magnetic force used. The maximum amplitude of the magnetic force output from the GA is 168.22 N while the result of the FEM simulation is 167 N which is very close to the GA value.

Table 6. Limitations for objective function

Parameter	Allowed domain	Parameter	Allowed domain
Air-gap width	0.2 mm	Width of the rings (w)	$3\ mm \le W \le 10\ mm$
Height of the rings (h)	$3\ mm \le h \le 10\ mm$	Magnetic force (F)	140 N ≤ F ≤ 170 N
Inner radius of outer ring (r)	24 mm ≤ r ≤ 26 mm	Force-to-cost ratio	Maximum in the considered domain

Table 7. Results of the optimization

Parameter	Value	Parameter	Value
Height	9.86 mm	Air-gap	0.2 mm
Width	5.6 mm	Magnetic force	168.22 N
Radius of outer ring	25.51 mm		

RESULT AND CONCLUSION

Passive Magnetic Bearing is an interesting topic for the industry due to its low cost. It is being applicable in machines and devices in which the precision of the coupling is not an important issue. A method was used to investigate the behavior of PMB considering most. FEM allowed achieving a precise investigation on PMB behavior.

Numerical analysis is performed to investigate the behavior of the most important parameters of PMB. Results of the numerical analysis are used as the input data for the genetic algorithm as the fitness and penalty functions and also as the first generation for GA.

Although magnetic levitation overcomes the drawbacks of the mechanical ball bearings, the force amplitude and the mechanical stiffness decrease in the passive type magnetic bearings. In order to compensate this deficiency, it is important to apply an appropriate volume of the magnet.

In order to achieve a better structure of PMB, it is suggested to add another magnet inside the previous structure of PMB. Adding one magnet improves the magnetic field and increases the magnetic force amplitude and consequently the magnetic stiffness. The total volume of PMB is not increased because the third ring is added inside of the PMB structure and also the mechanical stiffness will keep the PMB strictly in a fixed position and vibration and misalignment in PMB will be damped.

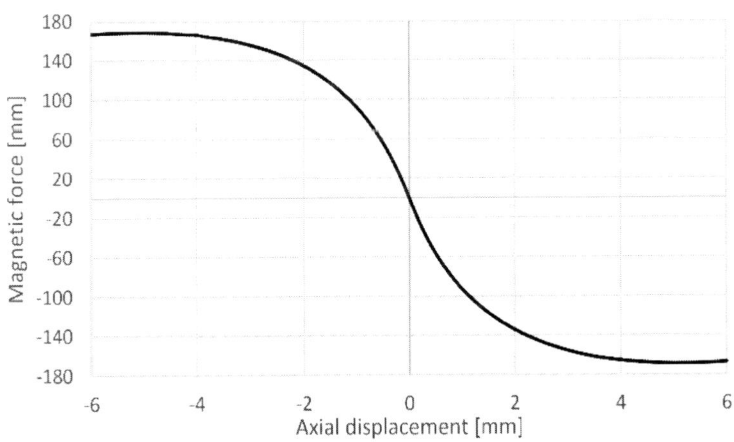

Figure 14. Force diagram versus axial displacement for the optimized dimension of PMB.

References

[1] Yanliang, X., Yueqin, D., Xiuhe, W. and Yu, K. Analysis of Hybrid Magnetic Bearing with a Permanent Magnet in the Rotor by FEM. *IEEE Trans. On Magn.* Vol. 42, no. 4, pp. 1363-1366, Apr. 2006.

[2] Eryong, H. and Kun, L. A Novel Structure for Low-Loss Radial Hybrid Magnetic Bearing. *IEEE Trans. On Magn.* Vol. 47, no. 12, pp. 4725-4733, Dec. 2011.

[3] Eryong, H. and Kun, L. Investigation of Axial Carrying Capacity of Radial Hybrid Magnetic Bearing. *IEEE Trans. On Magn.* Vol. 48, no. 1, pp. 38-46, Jan. 2012.

[4] Filatov, A. V. and Maslen, E. H., Passive Magnetic Bearing for Flywheel Energy Storage Systems. *IEEE Trans. On Magn.* Vol. 37, no. 6, pp. 3913-3924, Nov. 2001.

[5] Kimman, M. H., Langen, H. H. and Schmidt, R. H. M. A Miniature Milling Spindle with Active Magnetic Bearings. *Mechatronics*, Vol. 20, no. 2, pp. 224-235, Mar. 2010.

[6] Bangcheng, H. Shiqiang, Z., Qian, W. Xi, Y. Integral Design and Analysis of Passive Bearing and Active Radial Magnetic Bearing for Agile Satellite Application. *IEEE Trans. On Magn.* Vol. 48, no. 6, pp. 1959-1966, June 2012.

[7] Fleischer, E. and Hofmann, W. Application of Soft Magnetic Composites in Active Magnetic Bearings. in 37th Annual Conference on *IEEE Industrial Electronics Society*, IECON, pp. 1770-1775, Nov. 2011.

[8] AlizadehTir, M., Mirimani, S. M. and Marignetti, F. A Novel Structure of Passive Magnetic Bearing with Axial Magnetization. in 5th Power Electronics, Drive Systems and Technologies Conferences (*PEDSTC*), pp. 311-316, Feb. 2016.

[9] Jiancheng, F., Jinji, S., Yanliang, X. and Xi, W. A New Structure for Permanent-Magnet-Biased Axial Hybrid Magnetic Bearings. *IEEE Trans. On Magn.* Vol. 45, no. 12, pp. 5319-5325, Dec. 2009.

[10] Musolino, A., Rizzo, R. Passively Stabilized Magnetic Bearing. in 14th Biennial *IEEE Conference on Electromagnetic Field Computation* (CEFC), May 2010.

[11] Matsuda, K., Kijimoto, S. H. An Approach to Designing a Magnetic-Bearing System for Smaller Rotating Machines. *IEEE International Symposium on Industrial Electronics*, ISIE, pp. 1418-1422, July 2008.

[12] Beams, J. W. High Rotation Speed. *J. Applied physics*, 8:795-806, 1937.

[13] Asama, J., Asami, T., Imakawa, T., Chiba, A., Nakajima A. and AzizurRahman, M. Effects of Permanent-Magnet Passive Magnetic Bearing on a Two-Axis Actively Regulated Low-Speed Bearingless Motor. *IEEE Trans. On Energy Conversion*, Vol. 26, no. 1, pp. 46-54, March 2011.

[14] Impinna, F., Detoni, J. G., Amati, N. and Tonoli, A. Passive Magnetic Levitation of Rotors on Axial Electrodynamic Bearings. *IEEE Trans. On Magn.* Vol. 49, no. 1, pp. 599-608, Jan. 2013.

[15] Delamare, J., Rulliere, E., Yonnet, J. P. Classification and Synthesis of Permanent Magnet Bearing Configurations. *IEEE Trans. On Magn.* Vol. 31, no. 6, pp. 4190-4192, Nov. 1995.

[16] Ravaud, R., Lemarquand, G., Lemarquand, V., Depollier, C. Analytical calculation of the Magnetic Field Created by Permanent-Magnet Rings. *IEEE Trans. On Magnetics*, Vol. 44, no. 8, pp. 1982-1989, Aug. 2008.

[17] Babic, S. and Akyel, C. Improvement in the Analytical Calculation of the Magnetic Field Produced by Permanent Magnet Rings. *Progress in Electromagn. Res.C*, Vol. 5, pp. 71-82, 2008.

[18] Gotanda, H., Amano, R., Sugiura, T. Mode Coupling of a Flexible Rotor Supported by a Superconducting Magnetic Bearing Due to the Nonlinearity of Electromagnetic Force. *IEEE Trans. On Applied Superconductivity*, Vol. 21, no. 3, pp. 1481-1484, June 2011.

[19] Sun, J., Ju, Z., Peng, C., Le, Y., Ren, H. A Novel 4-DOF Hybrid Magnetic Bearing for DGMSCMG. *IEEE Trans. On Industrial Electronics*, Vol. 64, no. 3, pp. 2196-2204, March 2017.

[20] Qian, K. X., Xu, H. X. Gyro-Effect and Earnshaw's Theorem: Stable and Unstable Equilibrium for Rotary and Stationary Permanent Magnetic Levitators. in 2th International Conference on Bioinformatics and Biomedical Engineering, *ICBBE*, May 2008.

[21] Jungmayr, G., Marth, E., Amrhein, W., Berroth, H-J., Jeske, F. Analytical Stiffness Calculation for Permanent Magnetic Bearings with Soft Magnetic Materials. *IEEE Trans. On Magn.*, Vol. 50, no. 8, pp. 1-8, Aug. 2014.

[22] Ravaud, R., Lemarquand, G., Lemarquand, V. Force and Stiffness of Passive Magnetic Bearings Using Permanent Magnets. Part 1: Axial Magnetization. *IEEE Trans. On Magnetics*, Vol. 45, no. 7, pp. 2996-3002, July 2009.

[23] Ravaud, R., Lemarquand, G., Lemarquand, V. Force and Stiffness of Passive Magnetic Bearings Using Permanent Magnets. Part 2: Radial Magnetization. *IEEE Trans. On Magnetics*, Vol. 45, no. 9, pp. 3334-3342, Sep. 2009.

In: Advances in Engineering Research
Editor: Victoria M. Petrova
ISBN: 978-1-53614-218-1

Chapter 5

A UNIVERSAL APPROACH FOR THE MAGNETIC FORCE CALCULATION OF RADIAL AND AXIAL MAGNETIC BEARINGS

Ana N. Vučković*[*], *Dušan M. Vučković*, *Mirjana T. Perić* and *Nebojša B. Raičević
Faculty of Electronic Engineering, University Niš, Niš, Serbia

ABSTRACT

Because the magnetic bearings are contactless suspension devices, they have various rotating and translational applications. To control for the reliability of the device, it is very important to know the magnetic force interaction between the permanent magnets that are part of the device itself. This paper presents a universal solution for the magnetic force interaction calculation between arbitrarily magnetized permanent ring magnets. The configuration contains two permanent ring magnets that correspond to the axial or radial passive magnetic bearing

[*] Corresponding Author address: Email: ana.vuckovic@elfak.ni.ac.rs.

configuration. The type of the bearing depends on the permanent magnet's magnetization vector direction and the permanent magnet's position. Calculation of the magnetic force interaction is performed when the magnetization vectors of the magnets exhibit an inclination with respect to geometry. The inclination angles of the magnetization vectors determine if the bearing will be considered radial or axial. This paper proposes a simple and fast semi-analytical approach to overcome the disadvantages of numerical approaches (e.g., high computational cost) and the disadvantages of analytical methods which are not robust enough to provide the analytical solution for force. The proposed approach is based on the magnetic charges distribution and discretization technique. The results for the magnetic force interaction obtained using the proposed approach are given for different types of the magnetic bearings and compared using the Finite Element Method (FEM). The advantage of the approach presented is its simplicity and time efficiency. The proposed approach uses only complete elliptic integrals of the second kind and additional integrations are avoided. Also, the force expression obtained allows for the analysis of different types of passive magnetic bearings.

Keywords: magnetic bearing, magnetic charges, magnetic force, permanent ring magnet (PM)

INTRODUCTION

In modern society, there are many different applications for permanent magnets, not the least of which is their essential role in electrical devices. The magnet's material, magnetization, and dimensions affect the quality of the permanent magnet which eventually will be used in an electrical device or other application. Given the importance of permanent magnets in modern society, there is a growing need to dimensionalize and optimize these magnets. New calculation methods must be developed to fully realize the ability to dimensionalize and optimize magnets. This paper proposes a universal solution for the magnetic force interaction calculation.

There are two major application areas based on the shape of the magnet. One application area is for block magnets [1], [2] and the other is for cylindrical magnets [3], [9]. The level of complexity of the apparatus needed to determine the magnetic field value varies by whether the permanent magnet is a block or cylindrical magnet. It is more complex to determine the value for cylindrical magnets [10], [11].

The majority of engineering applications require several permanent ring magnets and the ability to calculate the actual magnetic force between them. These devices are basically exploiting the attraction or repulsion of the permanent ring magnets.

One good example is a magnetic bearing [7], [8]. These are contactless suspension devices with various rotating and translational applications [16]. The intention is to replace conventional mechanical bearings with the much cleaner technology of magnetic bearings. Further, magnetic bearings also provide levitation as well as centering and thrust control of the rotating parts. This can be similar to the way mechanical bearings perform. Being "contactless" not only avoids friction, but also allows them to work as an axial or radial bearing, depending on the permanent ring magnet's magnetization direction. This way, they control the position along the axis or its centering.

To have full control of the device, knowledge of the magnetic force interaction is an imperative. Potential use cases of magnetic bearings include flywheel energy storage systems for intermittent sources of clean, renewable energy; turbines; compressors; motors; generators; and pumps.

There are many different techniques for analyzing permanent magnet-based devices and approaches for calculating the magnetic field. This variety of applications has steered the research of authors into different directions. The calculation of the interaction forces that exist between permanent magnets is given in [5-11]. The general approach is to find simplified formulations of the interaction forces created by permanent magnets. The most common current method is to use Ampere's current model [5], [12] or the Coulombian approach [6-8].

The method presented in this paper is semi-analytical with the resulting expression provided in its native form. The elementary functions used are available in all programing languages. The final expression can be easily implemented in any compiler and should take no more than a second to run (on an average personal computer.) The minimal run time was achieved by using only complete elliptic integrals of the second kind thereby avoiding other integrations that require more CPU time.

The results presented in this paper are important for the modeling process itself as well as the actual production of different permanent magnet-based devices.

MODEL DESCRIPTION

Figure 1 shows the system of two arbitrarily magnetized permanent ring magnets with a misalignment in the axial magnetization of the permanent magnets.

The approach based on fictitious magnetic charges and discretization technique is used to calculate the magnetic force. Considering the magnetization vector direction of the upper magnet (Figure 1), there are axial $(\boldsymbol{M}_{1z})$ and radial $(\boldsymbol{M}_{1r})$ magnetization vector components.

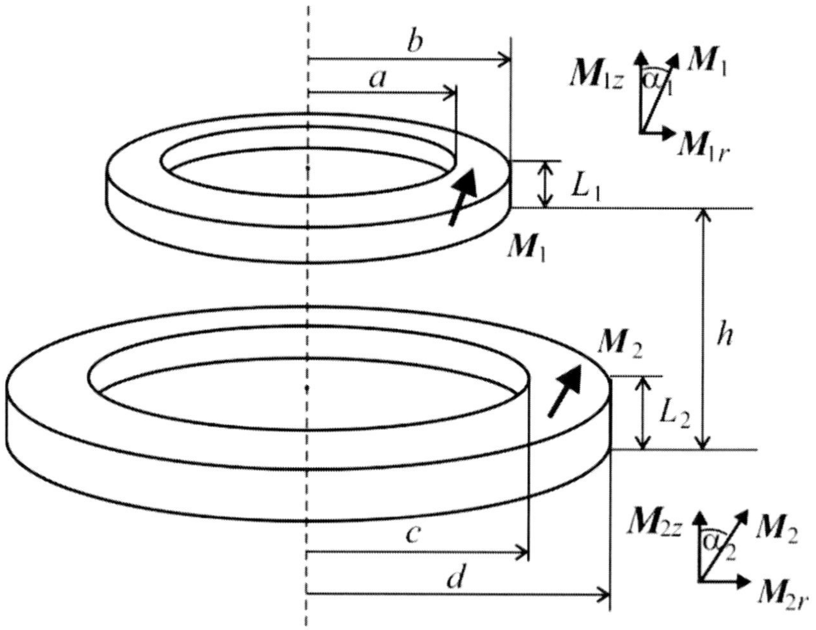

Figure 1. System of arbitrarily magnetized permanent ring magnets.

$$\boldsymbol{M}_1 = \boldsymbol{M}_{1r} + \boldsymbol{M}_{1z} = M_1 \sin\alpha_1 \cdot \hat{r} + M_1 \cos\alpha_1 \cdot \hat{z} \tag{1}$$

Because there is a magnetization vector component in axial direction, the fictitious magnetic charges exist on the bottom and the top bases of the magnet (Figure 2).

$$\begin{aligned} \eta_{mb1} &= \hat{n}_{b1} \cdot \boldsymbol{M}_{1z} = -M_{1z} = -M_1 \cos\alpha_1 \\ \eta_{mb2} &= \hat{n}_{b2} \cdot \boldsymbol{M}_{1z} = M_{1z} = M_1 \cos\alpha_1 \end{aligned} \tag{2}$$

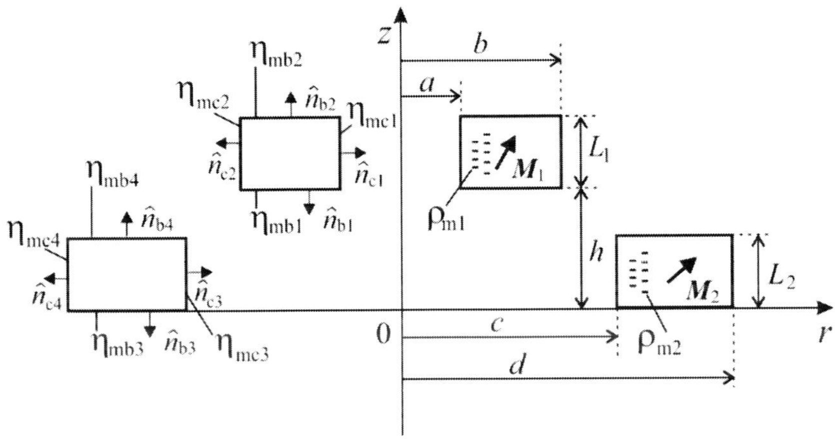

Figure 2. Distribution of surface and volume magnetization charges.

Because of the radial magnetization vector component, there are surface magnetization charges on the permanent ring magnet covers.

$$\begin{aligned} \eta_{mc1} &= \hat{n}_{c1} \cdot \boldsymbol{M}_{1r} = -M_{1r} = -M_1 \sin\alpha_1 \\ \eta_{mc2} &= \hat{n}_{c2} \cdot \boldsymbol{M}_{1r} = M_{1r} = M_1 \sin\alpha_1 \end{aligned} \tag{3}$$

The volume charges inside the magnet.

$$\rho_{m1} = -\nabla \cdot \boldsymbol{M}_1 = -M_{1r}/r = -M_1 \sin\alpha_1 / r \tag{4}$$

The magnetization vector of the second magnet can be presented using its components.

$$\boldsymbol{M}_2 = \boldsymbol{M}_{2r} + \boldsymbol{M}_{2z} = M_2 \sin\alpha_2 \cdot \hat{r} + M_2 \cos\alpha_2 \cdot \hat{z} \tag{5}$$

The surface magnetic charges of the lower magnet's bases are calculated as follows.

$$\begin{aligned} \eta_{\mathrm{mb3}} &= \hat{n}_{\mathrm{b3}} \cdot \boldsymbol{M}_{2z} = -M_{2z} = -M_2 \cos\alpha_2 \\ \eta_{\mathrm{mb4}} &= \hat{n}_{\mathrm{b4}} \cdot \boldsymbol{M}_{2z} = M_{2z} = M_2 \cos\alpha_2 \end{aligned} \tag{6}$$

The magnetic charges of the permanent magnet covers are calculated as follows.

$$\begin{aligned} \eta_{\mathrm{mc3}} &= \hat{n}_{\mathrm{c3}} \cdot \boldsymbol{M}_{2r} = -M_{2r} = -M_2 \sin\alpha_2 \\ \eta_{\mathrm{mc4}} &= \hat{n}_{\mathrm{c4}} \cdot \boldsymbol{M}_{2r} = M_{2r} = M_2 \sin\alpha_2 \end{aligned} \tag{7}$$

The volume charges of the lower magnet can be expressed as follows.

$$\rho_{\mathrm{m2}} = -\nabla \cdot \boldsymbol{M}_2 = -M_{2r}/r = -M_2 \sin\alpha_2 / r \tag{8}$$

Therefore, when it comes to permanent magnets magnetized in an arbitrary direction, the permanent ring magnets' covers, bases, and volume can be discretized into a system of circular loops loaded with different magnetic charges. Figure 2 presents the distribution of the permanent ring magnet's magnetic charges for the considered configuration.

To calculate the magnetic force that exists between two arbitrarily magnetized permanent ring magnets, one must first derive the magnetic force between the two circular loops loaded with different magnetic charges [13].

The expressions of the magnetic force components between two circular loops loaded with different magnetic charges (Figure 3) are as follows.

$$F_r = \mu_0 \frac{Q_{m1}Q_{m2}}{2\pi^2} \times \left(\frac{\left(r_m^2 - r_0^2 - (z_m - z_0)^2\right) E\left(\frac{\pi}{2}, k_0\right)}{2r_m \left((r_m - r_0)^2 + (z_m - z_0)^2\right)\sqrt{(r_m + r_0)^2 + (z_m - z_0)^2}} + \frac{K\left(\frac{\pi}{2}, k_0\right)}{2r_m \sqrt{(r_m + r_0)^2 + (z_m - z_0)^2}} \right) = 0 \tag{9}$$

$$F_z = \mu_0 \frac{Q_{m1}Q_{m2}}{2\pi^2} \times \frac{(z_m - z_0) E\left(\frac{\pi}{2}, k_0\right)}{\left((r_m - r_0)^2 + (z_m - z_0)^2\right)\sqrt{(r_m + r_0)^2 + (z_m - z_0)^2}} = \frac{\mu_0}{2\pi^2} Q_{m1}Q_{m2}F_{z_p}(r_0, r_m, z_0, z_m) \tag{10}$$

The elliptic integrals modulus is: $k_0^2 = \dfrac{4r_0 r_m}{(r_m + r_0)^2 + (z_m - z_0)^2}$

The interaction force between the two magnetized circular loops is the axial component of the force, as calculated by equation (10).

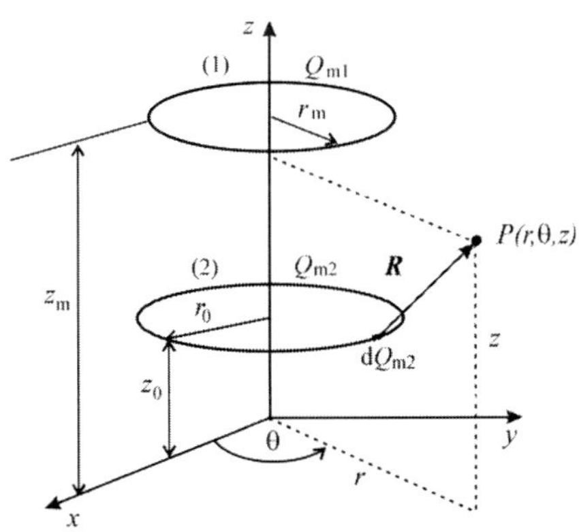

Figure 3. Two circular loops loaded with different magnetic charges.

The approach presented in the paper is based on discretizing:

- the permanent magnets' covers and bases into a system of $N_{\mathrm{s}}^{\mathrm{tot}}$ circular loops; and
- the volume of permanents magnets into system of $N_{\mathrm{v}}^{\mathrm{tot}}$ circular loops.

The discretization model is presented in Figure 4. Figure 4.a shows the distribution of all surface charges. Figure 4.b shows the volume segments.

The parameters of the system can be defined easily from Figure 4.

The parameters for the bases of the upper permanent magnet are shown in equations (11) and (12).

$$r_{1n} = a + \frac{2n_1 - 1}{2N_{\mathrm{b1}}}(b-a), \;\; z_{1n} = h, \;\; n_1 = 1,2,\ldots,N_{\mathrm{b1}} \tag{11}$$

$$r_{2n} = r_{1n}, z_{2n} = h + L_1 \tag{12}$$

The magnetic charges of the lower and upper bases are calculated as follows.

$$Q_{\mathrm{m}b_1} = Q_{\mathrm{m}b_{1n_1}} = -M_{1z} 2\pi r_{1n} \frac{b-a}{N_{\mathrm{b1}}}, n_1 = 1,2,\ldots,N_{\mathrm{b1}} \tag{13}$$

$$Q_{\mathrm{m}b_2} = Q_{\mathrm{m}b_{2n_1}} = -Q_{\mathrm{m}b_1} \tag{14}$$

N_{b1} is the number of discretization segments (circular loops) of each base of the upper PM.

The parameters of permanent magnet cover segments are calculated using equations (15) and (16).

$$r_{1m} = a, \;\; z_{1m} = h + \frac{2m_1 - 1}{2N_{\mathrm{c1}}} L_1, \;\; m_1 = 1,2,\ldots,N_{\mathrm{c1}} \tag{15}$$

$$r_{2m} = b,\ z_{2m} = z_{1m} \tag{16}$$

The magnetic charges of the inner and outer covers are as follows.

$$Q_{\mathrm{mc}_1} = Q_{\mathrm{mc}_{1m_1}} = -M_{1r}2\pi a\frac{L_1}{N_{\mathrm{c}1}}, m_1 = 1,2,\dots,N_{\mathrm{c}1} \tag{17}$$

$$Q_{\mathrm{mc}_2} = Q_{\mathrm{mc}_{2m_1}} = M_{1r}2\pi b\frac{L_1}{N_{\mathrm{c}1}}, m_1 = 1,2,\dots,N_{\mathrm{c}1} \tag{18}$$

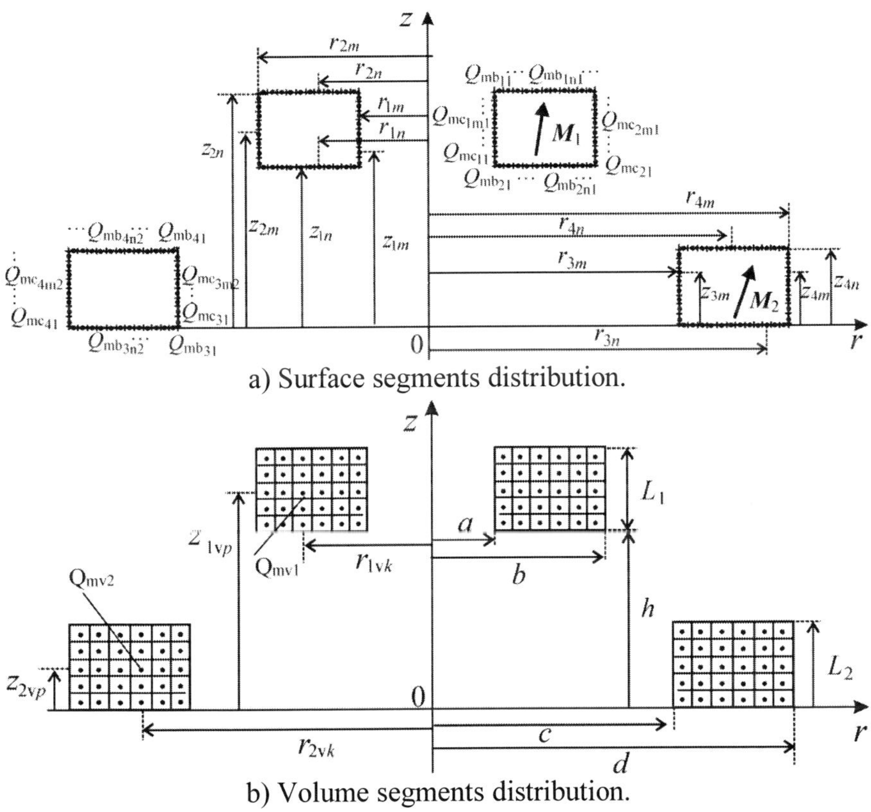

Figure 4. Discretization model.

$N_{\mathrm{c}1}$ is the number of the upper magnet's cover segments.

$N_{\mathrm{b}1}$ and $N_{\mathrm{c}1}$ are calculated from the initial number of surface segments, $N_{\mathrm{s}}^{\mathrm{tot}}$, and they depend on the permanent magnet's dimensions.

The volume segments parameters are calculated as follows.

$$z_{1\mathrm{vp}} = h + \frac{2p_1 - 1}{2N_{\mathrm{v}1}} L_1, \; p_1 = 1, 2, \ldots, N_{\mathrm{v}1} \tag{19}$$

$$r_{1\mathrm{v}k} = a + \frac{2k_1 - 1}{2N_{\mathrm{v}2}} (b - a), \; k_1 = 1, 2, \ldots, N_{\mathrm{v}2} \tag{20}$$

Their respective magnetic charges can be computed using equation (21).

$$Q_{\mathrm{mv}1} = Q_{\mathrm{mv}1\,j1} = -M_{1r} 2\pi L_1 \frac{b - a}{N_{\mathrm{v}1} N_{\mathrm{v}2}}, \; j_1 = 1,2,\ldots,N_{\mathrm{v}}^{\mathrm{tot}} \tag{21}$$

$N_{\mathrm{v}}^{\mathrm{tot}} = N_{\mathrm{v}1} \cdot N_{\mathrm{v}2}$ is the total number of volume segments of the upper permanent magnet.

Considering the discretization model, as shown in Figure 4, the following parameters can be defined for the lower permanent magnet.

$$r_{3n} = c + \frac{2n_2 - 1}{2N_{\mathrm{b}2}} (d - c), \; z_{3\mathrm{n}} = 0, \quad n_2 = 1, 2, \ldots, N_{\mathrm{b}2} \tag{22}$$

$$r_{4n} = r_{3n}, z_{4n} = L_2 \tag{23}$$

The magnetic charges of the lower and upper bases are computed as follows.

$$Q_{\mathrm{m}b_3} = Q_{\mathrm{m}b_{3n_2}} = -M_{2z} 2\pi r_{3n} \frac{d-c}{N_{\mathrm{b}2}}, n_2 = 1,2,\ldots,N_{\mathrm{b}2} \tag{24}$$

$$Q_{\mathrm{m}b_4} = Q_{\mathrm{m}b_{4n_2}} = -Q_{\mathrm{m}b_3} \tag{25}$$

The number of discretization segments for each base of the lower PM is N_{b2}.

The parameters of the permanent magnet cover segments are as follows.

$$r_{3m} = c, \; z_{3m} = \frac{2m_2 - 1}{2N_{c2}} L_2, \quad m_2 = 1, 2, \ldots, N_{c2} \tag{26}$$

$$r_{4m} = d, \; z_{4m} = z_{3m} \tag{27}$$

The magnetic charges of the inner and outer covers are calculated using equations (28) and (29).

$$Q_{mc_3} = Q_{mc_{3m_2}} = -M_{2r} 2\pi c \frac{L_2}{N_{c2}}, m_2 = 1,2, \ldots, N_{c2} \tag{28}$$

$$Q_{mc_4} = Q_{mc_{4m_2}} = M_{2r} 2\pi d \frac{L_2}{N_{c2}}, m_2 = 1,2, \ldots, N_{c2} \tag{29}$$

N_{c2} is the number of the lower permanent magnet cover segments.

N_{b2} and N_{c2} are calculated from the initial number of surface segments, N_s^{tot}, and they depend on the lower permanent magnet's dimensions.

The volume segments parameters can also be calculated.

$$z_{2vp} = \frac{2p_2 - 1}{2N_{v3}} L_2, \; p_2 = 1, 2, \ldots, N_{v3} \tag{30}$$

$$r_{2vk} = c + \frac{2k_2 - 1}{2N_{v4}} (d - c), \; k_2 = 1, 2, \ldots, N_{v4} \tag{31}$$

Also, their respective magnetic charges.

$$Q_{mv2} = Q_{mv2j2} = -M_{2r} 2\pi L_2 \frac{d - c}{N_{v3} N_{v4}}, \; j_2 = 1,2, \ldots, N_v^{tot} \tag{32}$$

$N_{\mathrm{v}}^{\mathrm{tot}} = N_{\mathrm{v}3} \cdot N_{\mathrm{v}4}$ is the total number of volume segments of the lower permanent magnet.

The magnetic force interaction between the two permanent ring magnets is obtained using the results for the magnetic force interaction between two circular loops, equation (10). It is achieved by summing the contribution of bases, covers, and volumes of the lower and upper permanent magnets by using the uniform discretization technique.

$$
\begin{aligned}
F_z = {} & F_{b1b3} + F_{b1b4} + F_{b2b3} + F_{b2b4} + F_{b1c3} + F_{b1c4} + F_{b2c3} + F_{b2c4} \\
& + F_{c1b3} + F_{c1b4} + F_{c2b3} + F_{c2b4} + F_{c1c3} + F_{c1c4} + F_{c2c3} \\
& + F_{c2c4} + F_{b1v2} + F_{b2v2} + F_{c1v2} + F_{c2v2} + F_{b3v1} + F_{b4v1} \\
& + F_{c3v1} + F_{c4v1} + F_{v1v2}
\end{aligned} \tag{33}
$$

$F_{b1b3} + F_{b1b4} + F_{b2b3} + F_{b2b4}$ is the force between the circular magnetized loops of the segments for the bases of both permanent magnets.

$F_{b1c3} + F_{b1c4} + F_{b2c3} + F_{b2c4}$ is the force between the cover surface segments of the lower magnet and the segments of the upper magnet's bases.

$F_{c1b3} + F_{c1b4} + F_{c2b3} + F_{c2b4}$ is the force between the cover surface segments of the upper magnet and the segments of the lower magnet's bases.

$F_{c1c3} + F_{c1c4} + F_{c2c3} + F_{c2c4}$ is the force between the covers segments of the permanent magnets.

$F_{b1v2} + F_{b2v2} + F_{c1v2} + F_{c2v2}$ is the force between the volume segments of the lower magnet and all the surface segments (cover and base) of the upper magnet.

$F_{b3v1} + F_{b4v1} + F_{c3v1} + F_{c4v1}$ is the force between the volume segments of the upper magnet and all the surface segments of the lower magnet.

F_{v1v2} is the force between the volume segments of the lower and the upper magnets.

Finally, the magnetic force interaction between the two permanent ring magnets has the following form.

$$
\begin{aligned}
F_z &= \frac{\mu_0}{2\pi^2}\left(\sum_{n_1=1}^{N_{b1}}\sum_{n_2=1}^{N_{b2}}\left(Q_{\mathrm{mb}_1}Q_{\mathrm{mb}_3}F_{\mathrm{zp}}(r_{3n},r_{1n},z_{3n},z_{1n})\right.\right.\\
&+Q_{\mathrm{mb}_1}Q_{\mathrm{mb}_4}F_{\mathrm{zp}}(r_{4n},r_{1n},z_{4n},z_{1n})+Q_{\mathrm{mb}_2}Q_{\mathrm{mb}_3}F_{\mathrm{zp}}(r_{3n},r_{2n},z_{3n},z_{2n})\\
&\left.+Q_{\mathrm{mb}_2}Q_{\mathrm{mb}_4}F_{\mathrm{zp}}(r_{4n},r_{2n},z_{4n},z_{2n})\right)\\
&+\sum_{n_1=1}^{N_{b1}}\sum_{m_2=1}^{N_{c2}}\left(Q_{\mathrm{mb}_1}Q_{\mathrm{mc}_3}F_{\mathrm{zp}}(r_{3m},r_{1n},z_{3m},z_{1n})\right.\\
&+Q_{\mathrm{mb}_1}Q_{\mathrm{mc}_4}F_{\mathrm{zp}}(r_{4m},r_{1n},z_{4m},z_{1n})+Q_{\mathrm{mb}_2}Q_{\mathrm{mc}_3}F_{\mathrm{zp}}(r_{3m},r_{2n},z_{3m},z_{2n})\\
&\left.+Q_{\mathrm{mb}_2}Q_{\mathrm{mc}_4}F_{\mathrm{zp}}(r_{4m},r_{2n},z_{4m},z_{2n})\right)\\
&+\sum_{m_1=1}^{N_{c1}}\sum_{n_2=1}^{N_{b2}}\left(Q_{\mathrm{mc}_1}Q_{\mathrm{mb}_3}F_{\mathrm{zp}}(r_{3n},r_{1m},z_{3n},z_{1m})\right.\\
&+Q_{\mathrm{mc}_1}Q_{\mathrm{mb}_4}F_{\mathrm{zp}}(r_{4n},r_{1m},z_{4n},z_{1m})+Q_{\mathrm{mc}_2}Q_{\mathrm{mb}_3}F_{\mathrm{zp}}(r_{3n},r_{2m},z_{3n},z_{2m})\\
&\left.+Q_{\mathrm{mc}_2}Q_{\mathrm{mb}_4}F_{\mathrm{zp}}(r_{4n},r_{2m},z_{4n},z_{2m})\right)\\
&+\sum_{m_1=1}^{N_{c1}}\sum_{m_2=1}^{N_{c2}}\left(Q_{\mathrm{mc}_1}Q_{\mathrm{mc}_3}F_{\mathrm{zp}}(r_{3m},r_{1m},z_{3m},z_{1m})\right.\\
&+Q_{\mathrm{mc}_1}Q_{\mathrm{mc}_4}F_{\mathrm{zp}}(r_{4m},r_{1m},z_{4m},z_{1m})+Q_{\mathrm{mc}_2}Q_{\mathrm{mc}_3}F_{\mathrm{zp}}(r_{3m},r_{2m},z_{3m},z_{2m})\\
&\left.+Q_{\mathrm{mc}_2}Q_{\mathrm{mc}_4}F_{\mathrm{zp}}(r_{4m},r_{2m},z_{4m},z_{2m})\right)\\
&+\sum_{n_1=1}^{N_{b1}}\sum_{p_2=1}^{N_{\mathrm{v}3}}\sum_{k_2=1}^{N_{\mathrm{v}4}}\left(Q_{\mathrm{mb}_1}Q_{\mathrm{mv}_2}F_{\mathrm{zp}}\left(r_{2\mathrm{v}k},r_{1n},z_{2\mathrm{v}p},z_{1n}\right)\right.\\
&\left.+Q_{\mathrm{mb}_2}Q_{\mathrm{mv}_2}F_{\mathrm{zp}}\left(r_{2\mathrm{v}k},r_{2n},z_{2\mathrm{v}p},z_{2n}\right)\right)\\
&+\sum_{m_1=1}^{N_{c1}}\sum_{p_2=1}^{N_{\mathrm{v}3}}\sum_{k_2=1}^{N_{\mathrm{v}4}}\left(Q_{\mathrm{mc}_1}Q_{\mathrm{mv}_2}F_{\mathrm{zp}}\left(r_{2\mathrm{v}k},r_{1m},z_{2\mathrm{v}p},z_{1m}\right)\right.\\
&\left.\left.+Q_{\mathrm{mc}_2}Q_{\mathrm{mv}_2}F_{\mathrm{zp}}\left(r_{2\mathrm{v}k},r_{2m},z_{2\mathrm{v}p},z_{2m}\right)\right)\right)
\end{aligned}
$$

$$+\sum_{n_2=1}^{N_{b2}}\sum_{p_1=1}^{N_{v1}}\sum_{k_1=1}^{N_{v2}}\Big(Q_{mb_3}Q_{mv_1}F_{zp}(r_{3n},r_{1vk},z_{3n},z_{1vp})$$

$$+Q_{mb_4}Q_{mv_1}F_{zp}(r_{4n},r_{1vk},z_{4n},z_{1vp})\Big)$$

$$+\sum_{m_2=1}^{N_{c2}}\sum_{p_1=1}^{N_{v1}}\sum_{k_1=1}^{N_{v2}}\Big(Q_{mc_3}Q_{mv_1}F_{zp}(r_{3m},r_{1vk},z_{3m},z_{1vp})$$

$$+Q_{mc_4}Q_{mv_1}F_{zp}(r_{4m},r_{1vk},z_{4m},z_{1vp})\Big)$$

$$+\sum_{p_1=1}^{N_{v1}}\sum_{k_1=1}^{N_{v2}}\sum_{p_2=1}^{N_{v3}}\sum_{k_2=1}^{N_{v4}}Q_{mv_1}Q_{mv_2}F_{zp}(r_{2vk},r_{1vk},z_{2vp},z_{1vp})\Bigg)$$

(34)

The function F_{z_p} is defined in equation (10).

Numerical Results

Using the expression obtained for the magnetic force between two arbitrarily magnetized permanent ring magnets, (34), the attraction or repulsion force can be calculated for any type of magnetic bearing. This can be achieved by selecting the permanent magnets' parameters and the angles of magnetization vector inclination.

Two types of magnetic bearings commonly used in electromagnetic devices, axial [15] and radial magnetic bearings [13], are considered in the paper.

Magnetic Bearing – Axial Magnetization

A magnetic bearing that is composed of two axially magnetized permanent ring magnets is analyzed first and presented in Figure 5. There is an assumption that the magnetization vectors of the magnets have the same intensity, ($M_1 = M_2 = M$). Using expression (34), the magnetic force of the system could be calculated easily when the angles of magnetization vectors are $\alpha_1 = \pi$ and $\alpha_2 = 0$.

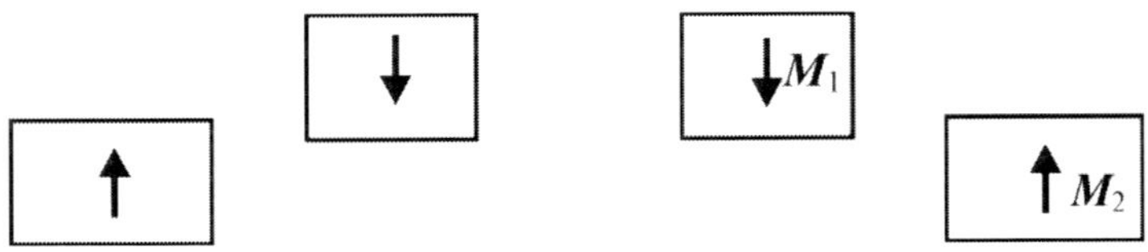

Figure 5. Axial magnetic bearing when $\alpha_1 = \pi$ and $\alpha_2 = 0$.

Highly accurate results can be obtained without using a large number of segments for the calculations. Figure 6 confirms this where the normalized magnetic force, $F_z^{\text{nor}} = F_z / \left(\mu_0 M^2 L_2^2\right)$, is presented versus the number of surface segments, N_s^{tot} . The configuration parameters used for the calculation are presented in Figure 6. The results obtained are compared with the results of the finite element method (FEMM 4.2 software) [4]. The relative error is less than 0.6% in the case when the number of surface segments is limited to N_s^{tot} =100. As the desired accuracy has been achieved, the total number of surface segments for both magnets, in the further calculations, is limited to 100. This creates a significant savings of calculation resources.

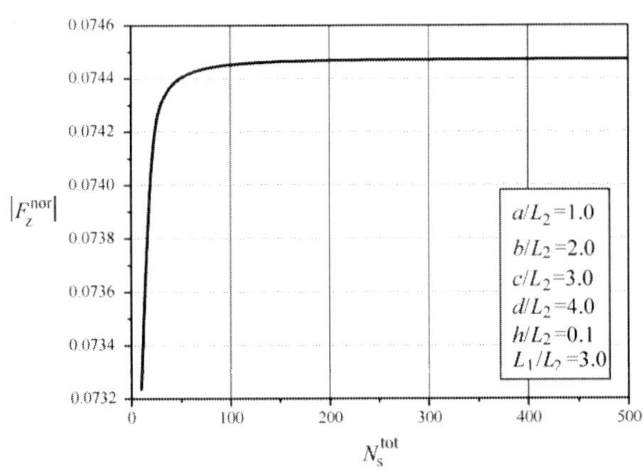

Figure 6. Normalized magnetic force versus the total number of surface segments.

The magnetic force's dependency on the variation of configuration parameters is presented graphically. Normalized magnetic force $F_z^{\mathrm{nor}} = F_z / (\mu_0 M^2 L_2^2)$ versus the axial displacement of the upper permanent magnet h/L_2 for different air gap dimension $\delta / L_2 = c/L_2 - b/L_2$ is presented in the Figure 7. Configuration parameters are $b/L_2 - a/L_2 = 1.0$, $c/L_2 = 2, d/L_2 = 3$ and $L_1 / L_2 = 1$.

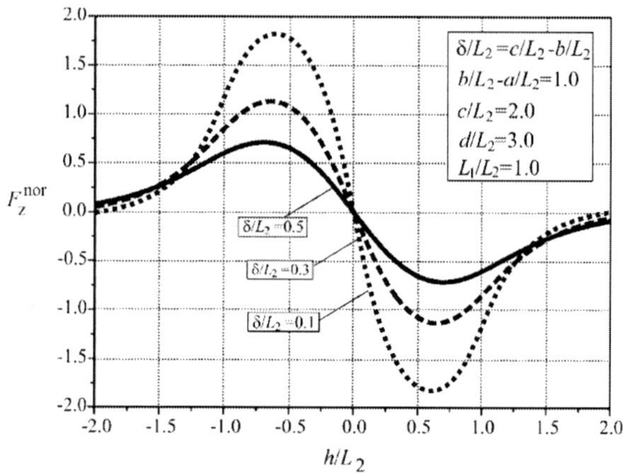

Figure 7. Normalized magnetic force versus ratio h/L_2 for different air gap dimension in case of axial magnetization of permanent magnets.

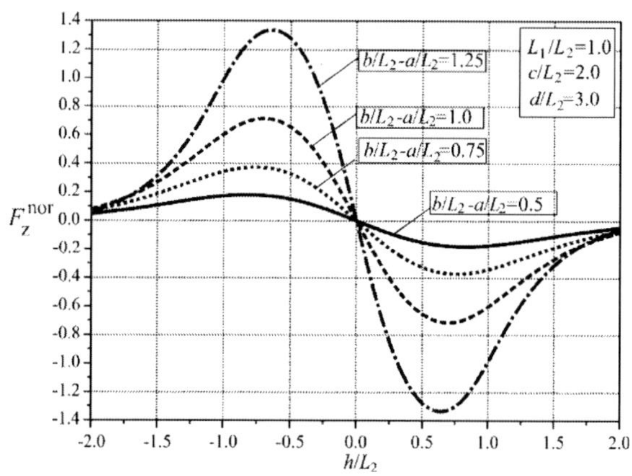

Figure 8. Normalized magnetic force versus the ratio, h/L_2, for different widths of the upper permanent magnet.

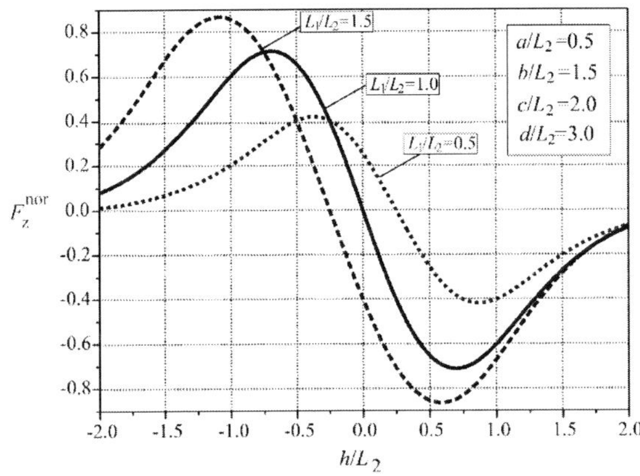

Figure 9. Normalized magnetic force versus the ratio, h/L_2, for different heights of permanent magnets.

Figure 8 shows the magnetic force interaction versus the ratio, h/L_2, for a different width of the upper permanent magnet, $b/L_2 - a/L_2$, for $c/L_2 = 2, d/L_2 = 3$, $L_1/L_2 = 1$ as the permanent magnet parameters.

Figure 9 shows the normalized magnetic force versus the axial displacement of the upper permanent magnet, h/L_2, for a different ratio of the permanent magnets' heights, L_1/L_2. The normalized force is plotted for the configuration parameters: $a/L_2 = 0.5, b/L_2 = 1.5$, $c/L_2 = 2.0, d/L_2 = 3.0$.

Considering the previous two figures, it is obvious that with an incremental increase of width and height of the upper magnet, the axial force raises. Besides, it is very important to find the balance in dimensions of permanent magnets because the costs of the magnets have to be taken in to account.

Magnetic Bearing – Radial Magnetization

The second configuration analyzed is a system of two radially magnetized permanent ring magnets made of the same material and magnetized uniformly along the radial direction with the same intensity

$(\boldsymbol{M}_1 = \boldsymbol{M}_2 = -M\cdot\hat{z})$, as shown in Figure 10. For the magnetic force calculation, expression (34) is used and the angles of the magnetization vectors misalignment are $\alpha_1 = -\pi/2$ and $\alpha_2 = -\pi/2$. After testing the convergence of the results, it is concluded that the appropriate accuracy is reached with $N_\mathrm{S}^\mathrm{tot} = N_\mathrm{V}^\mathrm{tot} = 100$ surface and volume segments.

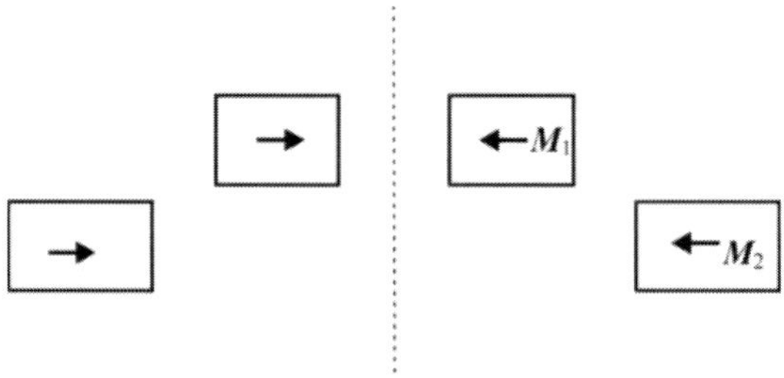

Figure 10. Radial magnetic bearing when $\alpha_1 = -\pi/2$ and $\alpha_2 = -\pi/2$.

The comparative results for the normalized magnetic force interaction, $F_z^{\mathrm{nor}} = F_z/\left(\mu_0 M^2 L_2^2\right)$, versus the normalized radii of the upper permanent ring magnet, a/L_2 and b/L_2, are obtained using the presented approach and finite element method (FEMM 4.2 software). These results are shown in Table 1 for the parameters $c/L_2 = 2.0, d/L_2 = 3.0$, $L_1/L_2 = 1.0$ and $h/L_2 = 1.5$. The force between the magnets could be attraction or repulsive, depending on the permanent magnets' dimensions and position.

Table 1. Comparative results for normalized magnetic force interaction versus a/L_2 and b/L_2

a/L_2	b/L_2	F_z^{nor}	F_z^{nor} (FEM)
0.5	1.5	-0.111398	-0.111609
1.0	2.0	0.046574	0.046183
1.5	2.5	0.817334	0.816931
2.0	3.0	1.499130	1.498061
2.5	3.5	0.946663	0.945968
3.0	4.0	-0.050398	-0.051031
3.5	4.5	-0.398905	-0.399117
4.0	5.0	-0.345845	-0.345966

Figure 11 presents the normalized magnetic force interaction versus the axial displacement of the upper permanent magnet, h/L_2, for different air gap dimensions, $\delta/L_2 = c/L_2 - b/L_2$. The parameters of the configuration are $b/L_2 - a/L_2 = 1.0, c/L_2 = 2.0, d/L_2 = 3.0$ and $L_1/L_2 = 1.0$.

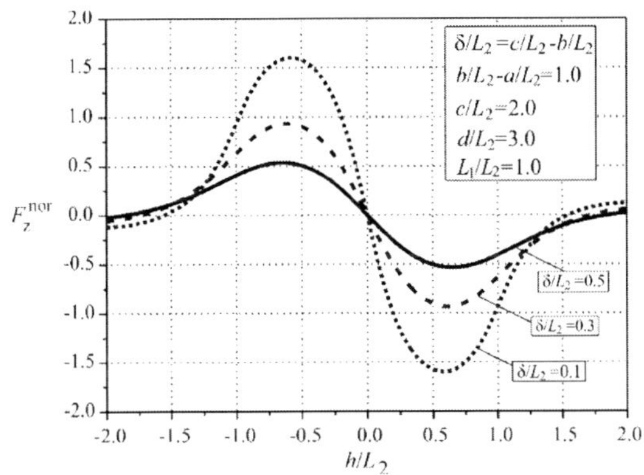

Figure 11. Normalized magnetic force versus axial displacement, h/L_2, for different air gap dimensions $\delta/L_2 = c/L_2 - b/L_2$.

As expected, the force increases as the air gap between the permanent ring magnets decreases. This is also demonstrated by the Figure 12, where the normalized interaction force versus the air gap dimension is presented.

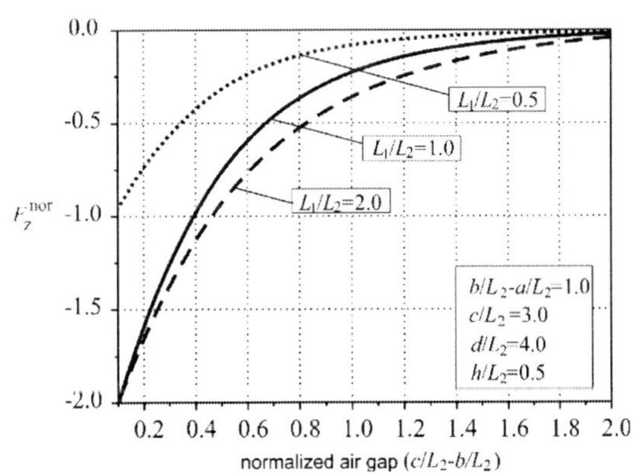

Figure 12. Normalized axial magnetic force versus air gap dimensions $\delta/L_2 = c/L_2 - b/L_2$.

Figure 13 and Figure 14 show the normalized axial magnetic force versus the ratio, h/L_2, for a different width and height of the upper permanent magnet. Considering the previous two figures, the absolute value of the axial force rises as the normalised width and height of the upper magnet rise, as expected. The balance of the dimensions of the permanent magnets must be found because the cost of the magnets should be taken into account.

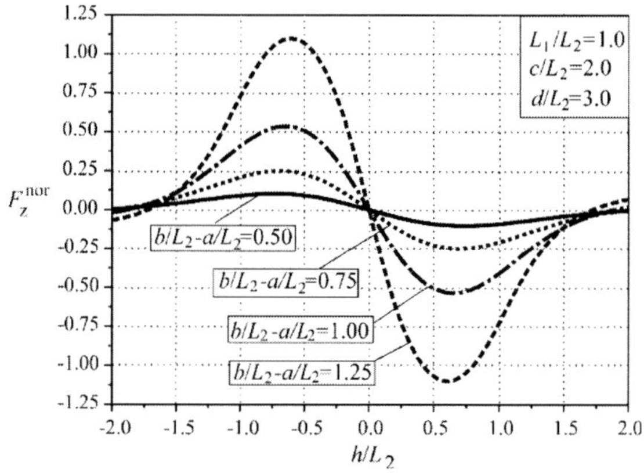

Figure 13. Normalized magnetic force versus axial displacement, h/L_2, for different widths of the upper permanent magnet, $b/L_2 - a/L_2$.

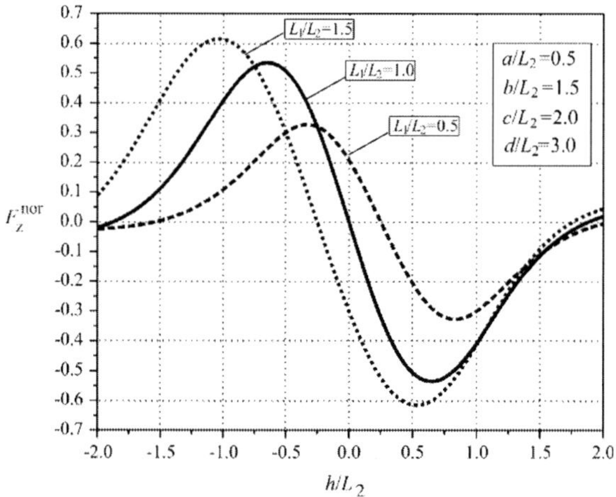

Figure 14. Normalized magnetic force versus the ratio, h/L_2, for different ratios of permanent magnets' heights, L_1/L_2.

The influence of the permanent magnets' cross section on the absolute value of the normalized magnetic force is presented in Figure 15. The parameters of each configuration are presented in the figure. Therefore, the expression (34) allows for the force to be analyzed for any type of passive magnetic bearing, regardless of the values of the parameters.

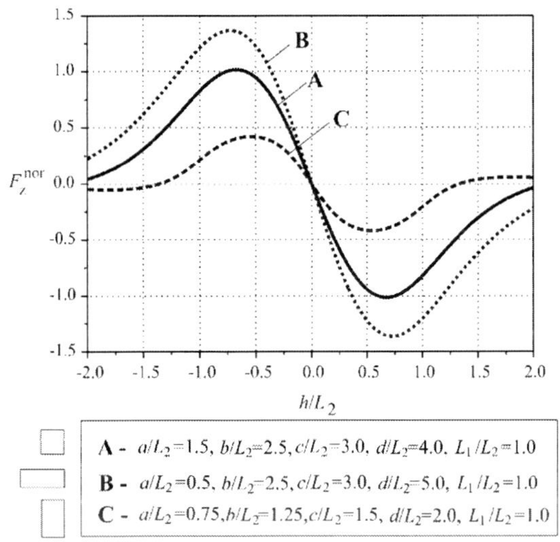

Figure 15. Normalized axial magnetic force versus the ratio, h/L_2, for different cross sections of permanent magnets.

CONCLUSION

The method based on the magnetic charges distribution, along with discretization technique, is used to find the universal solution for magnetic force interaction calculation between arbitrarily magnetized permanent ring magnets. The expression obtained allows the force calculation of different magnetic bearing configurations by selecting inclination angles of the magnetization vector.

The assumption is that both magnets are made of the same material and magnetized in an arbitrary direction with the same intensity. The derived algorithm is implemented easily in any standard computer environment and enables rapid parametric studies of the magnetic force for any type of passive magnetic bearing. The results of the presented approach are successfully confirmed using FEMM 4.2 software. Computational time can be saved by reducing the number of segments. It is confirmed that a large number of segments is not a prerequisite for the desired level accuracy to be achieved.

A computer with an Intel Core i5-2320 CPU Quad Core at 3GHz, 8GB RAM, and a 64-bit operating system was used to calculate the magnetic force interaction using the presented approach (for the aforementioned parameters and $N_{\mathrm{s}}^{\mathrm{tot}} = N_{\mathrm{v}}^{\mathrm{tot}} = 100$). Using this configuration, each calculation took about 0.76 seconds. Using the same computer and FEMM 4.2 software to calculate the magnetic force interaction, the observed execution time was 5.6 minutes for approximately 2 million finite elements. The results obtained can have a significant impact on the design of magnetic bearings, which are a primary component of many magnetic devices. By using only elliptical integral of second kind (along with elementary functions,) the final expression yields simplicity and efficiency when compared to contemporary methods. This method also allows field and force analysis in relation to the magnetic parameters, which is important for control of the devices containing permanent magnets.

ACKNOWLEDGMENTS

The presented work was partly supported by the Serbian Ministry of Education and Science in the frame of the project TR 33008.

REFERENCES

[1] Akoun, G., Yonnet, J. P., (1984). 3D analytical calculation of the forces exerted between two cuboidal magnets. *IEEE Transactions on magnetics. 20(5)*, 1962-1964.

[2] Beleggia, M., Vokoun, D., Graef, M. D., (2012). Forces between a permanent magnet and a soft magnetic plate. *IEEE Magnetics letters, 3*.

[3] Furlani, E. P., Reznik, S., Kroll, A., (1995). A three-dimensional field solution for radially polarized cylinders. *IEEE Transactions on magnetics. 31(1)*, 844–851.

[4] *Meeker, Software package FEMM 4.2*. Download (accessed 2 March 2007). Available on-line at http://www.femm.info/wiki/.

[5] Mladenović, A. N., Aleksić, S. R., Ilić, S. S., (2010). *Levitation force calculation for permanent magnet bearings using Ampere's currents*. Paper presented at the 14th International IGTE Symposium on Numerical Field Calculation in Electrical Engineering (pp. 149-153). Graz, Austria.

[6] Rakotoarison, H. L., Yonnet, J. P., Delinchant, B., (2007). Using Coulombian approach for modeling scalar potential and magnetic field of a permanent magnet with radial polarization. *IEEE Transactions on magnetics. 43(4)*, 1261-1264.

[7] Ravaud, R., Lemarquand, G., Lemarquand, V., (2009). Force and stiffness of passive magnetic bearings using permanent magnets. Part 1: Axial magnetization. *IEEE Transactions on Magnetics. 45(7)*, 2996-3002.

[8] Ravaud, R., Lemarquand, G., Lemarquand, V., (2009). Force and stiffness of passive magnetic bearings using permanent magnets. Part 2: Radial magnetization. *IEEE Transactions on Magnetics. 45(9)*, 3334-3342.

[9] Ravaud, R., Lemarquand, G., Babic, S., Lemarquand, V., Akeyel, C., (2010). Cylindrical magnets and coils: fields, forces and inductances. *IEEE Transactions on Magnetics. 46(9)*, 3585-3590.

[10] Robertson, W. S., Cazzolato, B. S., Zander, A. C., (2011). A Simplified force equation for coaxial cylindrical magnets and thin coils. *IEEE Transactions on magnetics. 47(8)*, 2045–2049.

[11] Vokoun, D., Beleggia, M., Heller, L., Sittner, P., (2009). Magnetostatic interactions and forces between cylindrical permanent

magnets, *Journal of magnetism and magnetic materials. 321(2)*, 3758-3763.

[12] Vučković, A. N., Ilić, S. S., Aleksić, S. R., (2012). Magnetic force interaction calculation of ring permanent magnets using Ampere's microscopic surface currents and discretization technique. *Electromagnetics. 32(2)*, 117-134.

[13] Vučković, A. N., Raičević, N. B., Ilić, S. S., Aleksić, S. R., (2013). Magnetic force interaction calculation of radial passive magnetic bearing using magnetization charges and discretization technique. *International journal of applied electromagnetics and mechanics. 43(4)*, 311-323.

[14] Vučković, A. N., Ilić, S.S., Aleksić, S. R., (2013). Magnetic force interaction calculation of permanent magnets using magnetization charges and discretization technique. *Electromagnetics. 33(6*), 421-436.

[15] Vučković, A. N., Aleksić, S. R., Ilić, S. S., Tatarenko, A., Petrov, R., Bichurin M., (2013). Attraction magnetic force calculation of axial passive magnetic bearing. *Electrical and electronic engineering. 3(2)*, 43-48.

[16] Yonnet, J. P., (1978). Passive magnetic bearings with permanent magnets. *IEEE Transactions on magnetics. 14(5*), 803–805.

INDEX

A

B

C

D

E

F

G

H

I

K

L

M

V